D0179853

THE

# SCIENTIFIC ENDEAVOR

*A Primer on Scientific Principles and Practice*

*Jeffrey A. Lee*

Texas Tech University

*an imprint of*
ADDISON WESLEY LONGMAN, INC.
*San Francisco • Reading, Massachusetts • New York • Harlow, England*
*Don Mills, Ontario • Sydney • Mexico City • Madrid • Amsterdam*

*For Linda*

Executive Editor: Erin Mulligan
Sponsoring Editor: Lynn Cox
Project Editor: Virginia Simione-Jutson
Managing Editor: Laura Kenney
Text Designer and Compositor: BookMatters
Illustrator: Bill Nelson
Cover Designer: Yvo Riezebos
Cover image: © copyright 1999 Photodisc, Inc.

*Library of Congress Cataloging-in-Publication Data*
Lee, Jeffrey A., 1956–
The scientific endeavor : a primer on scientific principles
and practice / Jeffrey A. Lee
p. cm.
Includes index.
ISBN 0-8053-4596-5
1. Science Popular works. I. Title.
Q162.L375 1999
500—dc21
8 9 10—MA—03

An imprint of Addison Wesley Longman, Inc.
1301 Sansome Street
San Francisco, CA 94111

# CONTENTS

# PREFACE

The purpose of *The Scientific Endeavor* is to present the basic underpinnings of science in a simplified way. It provides a generic view of science, not one tied to any specific discipline. Readers will not learn much chemistry or economics or geology, for example, but they will learn about common themes that unite those fields under the category of science. Thousands of books are available on the specific disciplines of science, and many exist on various aspects of science in general, such as philosophy of science and scientific misconduct. *The Scientific Endeavor* introduces all the major topics of science in a brief, readable format.

Who am I and why did I write this book? I started graduate school nearly twenty years ago, and I have been doing and teaching science since then. I am a geographer, which has been an advantage because the realm of geography is quite broad and shares subjects and methods with most other scientific disciplines. Much of my research has dealt with various aspects of soil erosion by wind, giving me the opportunity to do both field and laboratory studies and to work with modelers, geologists, meteorologists, pedologists, political scientists, engineers, and economists. I have taught alongside physicists, chemists, and biologists, and this, too, has helped me understand their perspectives on science.

In teaching science courses, it became clear to me that we cover the details of a particular discipline, but rarely do students come out of our classes with a general comprehension of science itself. All students should understand the nature of science as part of becoming an educated person, and students majoring in the sciences need to be exposed to scientific principles early in their college education. I wrote this book with two purposes: the first is to provide the information all educated people should know about how science works; the second is to give students who intend to pursue careers in science or science teaching the basics of what scientists ought to know about science. Typically, the topics covered here are too general to be included in most discipline-based courses. Unfortunately, the result is that this information is gotten in a piecemeal fashion, if at all, by students; hence, the need for this book, which complements traditional science texts.

I see *The Scientific Endeavor* as a fairly conventional view of science, shared by most scientists, though others may disagree. In fact, there is much here with which to disagree, especially by those experienced in science. Space limitations do not allow all sides of every issue to be covered, and I have not attempted to do so. Such is the nature of introductory textbooks.

I thank the following people for reading and commenting on one or more chapters: Linda Jones, Gary Elbow, Mark McGinley, David Lamp, Tom Gill, and Kevin Mulligan, all of Texas Tech University; Bruce Rhoads, University of Illinois, Urbana-Champaign; Art Strahler, Columbia University; John Stout, Ted Zobeck, and James Mahan, U.S.D.A. Agricultural Research Service, Texas; Len Finegold, Drexel University; Evan Moffett, U.S. Census Bureau; Ted Hermary, McGill University; Richard Paul, Sonoma State University; and John Fioretta, an old friend representing the nonscientist public.

The publisher's reviewers were also most helpful: Aaron Cassill, University of Texas at San Antonio; David E. Fulford, Edinboro University of Pennsylvania; Mary D. Healy, Springfield College; Paul E. Hertz, Barnard College; Thomas M. Iliffe, Texas A&M University at Galveston; and Mike Seeds, Franklin & Marshall College.

I especially thank my student reviewers at Texas Tech University who commented on drafts of all the chapters: Cheryl Carroll, Jim Childress, Brennan Cook, Jeremy Gott, Rusty Hatchett, Greg Lumen, Matt Mitchell, Robyn Phillips, Jonathan Smither, Lauren Spears, Lee Voss, Stacy Baggerly, David Berry, Robie Burns, Jami Carothers, Andrea Friedrich, Lindsay Griffin, Emily Johansen, Marissa Kramer, Jennifer Latta, Amy McBride, Kathleen McMurray, Mitsi Pair, Jeff Stayton, Amy Jo Chilton, Robert Mooney, and Courtnie Smith.

And finally, I thank the staff of the Texas Tech University library. Long live Interlibrary Loan!

*Jeff Lee*
j.lee@ttu.edu

# CHAPTER I

# INTRODUCTION

Scientific habits of mind can help people in every walk of life to deal sensibly with problems that often involve evidence, quantitative considerations, logical arguments, and uncertainty; without the ability to think critically and independently, citizens are easy prey to dogmatists, flimflam artists, and purveyors of simple solutions to complex problems.

*F. James Rutherford and Andrew Ahlgren, 1990*[1]

## WHAT IS SCIENCE?

The question "What is science?" does not have a simple answer. The word *science* conjures up different images for different people. People not familiar with science may think of it as unpronounceable words and long numbers, or guys with white lab coats and poor social skills, or something that may destroy the world, or something that may make the world a wonderful place to live. None of these is an accurate image of science. Even scientists, though, have trouble agreeing on a definition. Arthur Strahler has given a definition that is useful: "Science is the acquisition of reliable but not infallible knowledge of the real world, including explanations of the phenomena."[2] Science is a way of learning about things; it is a process we use to improve our understanding of the universe and all that is in it. This book is an introduction to what science is, how it is done, and finally, what it is not.

Knowledge comes in many forms, though, and science is only one approach to its acquisition. Science is among a group of approaches where understanding is gained through research. This group also includes the humanities, mathematics, and logic. The humanities, such as literature, art, and much of history, are concerned with artistic phenomena and the mostly qualitative study of human affairs. Mathematics and logic are closely related to science but are not necessarily based on real-world phenomena; they seek to explain the outcomes of situations, but the situations do not have to exist in reality. Other aspects of knowledge are called "belief fields," such as religion, ethics, morality, and political ideology. From a highly simplified perspective, belief fields are based on faith in what is right and what is wrong, or good and bad. Most relevant to this discussion, belief fields are not necessarily based on physical reality.[3] This book is concerned with science as a way of acquiring knowledge, and the other approaches will rarely be mentioned. The reader, however, should not assume that other approaches are inferior to science because they are not discussed. They simply deal with other types of phenomena.

The realm of science includes anything observable in the universe. As Strahler has stated, "Science gathers, processes, classifies, analyzes and stores information on anything and everything observable in the universe."[4] For something to be a subject of scientific inquiry it must be *measurable* in one way or another. Science is, therefore, *empirical* in that it relies on observation and experience. Value judgments are not scientifically measurable. You cannot measure "beauty" or "goodness," for example, because each person has a different interpretation of them.

While ethics, art, and religion are not scientific pursuits, sometimes science can assist these disciplines. For example, science can aid in the restoration of medieval stained glass windows[5] and in improving the acoustics of concert halls.[6] Another example is research on the Shroud of Turin, which was believed by many to be the cloth Jesus was wrapped in after the crucifixion, and the image of a face on the shroud was believed to be his. Researchers using scientific dating techniques have shown that the cloth is from the thirteenth or fourteenth century, and so did not exist at the time of Jesus's death.[7]

Of course, art and religion have aided science as well. Projective geometry was originally developed by Renaissance painters, but it later was used in the development of the theory of relativity.[8] The idea of camouflage, an important concept in evolutionary biology, was first developed by painter Abbott Thayer.[9] On a more general level, careful observation is a crucial skill for scientists, and one which can be developed through training in the arts.[10] As to religion, the Old Testament, for example, is a useful source of information on Middle Eastern archeology and environmental history.[11]

## AREAS OF SCIENCE

The major areas of science are the physical, biological, behavioral, and earth sciences. The *physical sciences* are physics and chemistry, which are concerned with the fundamental nature of matter and energy. Closely controlled experiments, usually done in laboratories, are the main type of research. The *biological sciences* deal with living things, their component parts, and the interactions between organisms, while *behavioral sciences* mainly involve the study of human behavior and organization, both in individuals (psychology) and in groups (the social sciences). The *earth sciences* are primarily concerned with nonliving matter on Earth and elsewhere in the universe. The biological, behavioral, and earth sciences rely heavily on observation outside the laboratory.[12]

It is important to appreciate that many scientific disciplines cross the boundaries of these categories. Anthropology, for example, is both a social and biological science; astronomy involves both earth and physical sciences; and geography incorporates aspects of behavioral, biological, and earth sciences. It is also important to understand that the disciplines we encounter at the university are, to some extent, artificially determined. This was well stated in 1887 by geographer Halford Mackinder, who wrote, "The truth of the matter is that the bounds of all the sciences must naturally be compromises. Knowledge . . . is one. Its division into subjects is a concession to human weakness."[13] There is a broad spectrum of subject matter, and sometimes science is hindered by the arbitrary categorization of "fields of study." Environmental problems, for example, need to be studied from a variety of perspectives. To understand river flooding, scientists have to incorporate knowledge of vegetation, land use, meteorology, and geology. Few social problems can be studied effectively without incorporating cultural attitudes, economics, and political conditions. To keep within the boundaries of one discipline is quite limiting and often hinders scientific progress. *Interdisciplinary studies* are becoming more common in science, and the boundaries of disciplines are a bit more flexible. In other words, science works best when the fields of study have lousy fences.

Keep in mind that disciplines classified as "sciences" may have nonscientific components. This is especially true in the social sciences, where much work is scientific but researchers often venture from the "What is . . ." questions of science to the "What should be . . ." questions of belief fields. Studying the causes of poverty, for example, can be scientific, while suggesting the morally best approach to reducing poverty is not. Much work better categorized as in the humanities is done by people in the social sciences, such as anthropologists using the methods of literary criticism to gain insight into a culture. Disciplines concerned with environmental problems, such as ecology, also have researchers who cross over into nonscientific research. On the other hand, while history usually is classified as a humanity, historical research sometimes is done as a social science. As long as the distinction is acknowledged between the science and nonscience areas of a discipline, this interaction can lead to a healthy exchange of ideas.

## BASIC AND APPLIED RESEARCH

While the subject matter of science can be divided into its various disciplines, scientific research can be categorized as either basic or applied. *Basic research* (sometimes called *pure science*) is driven by the curiosity of the scientist, who simply wants to learn more about a particular phenomenon. *Applied research*, on the other hand, is done to solve a specific problem and often leads to new or improved *technology*, which is the practical application of scientific knowledge. Studies of nuclear physics are done to understand the fundamental nature of matter, which fits under the category of basic research. Knowledge of nuclear physics also has been used in research into the development of nuclear weapons, nuclear medicine, and nuclear energy, which are areas of applied science. The specific weapons, medical devices, and reactors built using the knowledge gained are the resulting technologies. Similarly, ecological studies done to improve knowledge of the interactions of species with each other and the physical environment can be called basic research, while a study done to determine the most effective way to save an endangered species is called applied research. A governmental policy to protect the endangered species can be considered the resulting technology.

Basic and applied research are intertwined. Electronic devices, perhaps the defining technology of the late twentieth century, could not exist without knowledge discovered in the bizarre and esoteric basic research field of quantum mechanics. Applied research relies heavily on basic research to provide the underlying ideas that then are used to solve a problem. In turn, basic research affects society through the results of related applied research.

Whole scientific disciplines are organized around applied research, such as engineering, agronomy, urban planning, and medical research. Many of these applied-science disciplines have associated professions whose members, while not personally involved in scientific research, use scientific knowledge to solve problems. Examples are medical doctors, civil engineers, and economic analysts.

Politicians and others sometimes suggest that all of our scientific research should be directed at solving important problems rather than satisfying scientific curiosity. Most scientists disagree with this idea on the grounds that it is shortsighted. Some knowledge gained from basic research can be applied to improve economic, social, or environmental conditions, while the results of other basic research may never find a practical use. The problem is that no one can predict which research areas will lead to useful applications. (Or, as Carl Sagan puts it, "[w]e are rarely smart enough to set about on purpose making the discoveries that will drive our economy and safeguard our lives."[14]) Early in the nineteenth century, for example, few people could see any practical use for electricity; it was completely in the realm of basic research. This is shown by the story of Michael Faraday, who gave a tour of his laboratory to the British Prime Minister, Sir Robert Peel. Peel was interested in a newly invented dynamo, a device for generating electricity. As L. Pearce Williams tells the story, "Pointing to this odd machine, he inquired of what use it was. Faraday is said to have replied, 'I know not, but I wager that one day your government will tax it.' It was to be long after Sir Robert and Faraday had died that this prophecy came true for it was not until after the 1880's that the electrical power industry became large enough to attract the eye of the Treasury."[15] Most scientists feel that there must be a healthy balance between basic and applied research—basic research to provide general knowledge and applied research to use that knowledge, ideally, for the betterment of humankind.

---

Maxwell wasn't thinking of radio, radar, and television when he first scratched out the fundamental equations of electromagnetism; Newton wasn't dreaming of space flight or communication satellites when he first understood the motion of the Moon; Roentgen wasn't contemplating medical diagnosis when he investigated a penetrating radiation so mysterious he called it "X-rays"; Curie wasn't thinking of cancer therapy when she painstakingly extracted minute amounts of radium from tons of pitchblende; Fleming wasn't planning on saving the lives of millions with antibiotics when he noticed a circle free of bacteria around a growth of mold; Watson and Crick weren't imagining the cure of genetic diseases when they puzzled over the X-ray diffractometry of DNA; Rowland and Molina weren't planning to implicate CFCs in ozone depletion when they began studying the role of halogens in stratospheric photochemistry.

*Carl Sagan, 1995*[16]

---

The distinction between basic and applied research is not as sharp as some believe. Much fundamental knowledge has been gained from studies initially done to solve a problem. Engineering studies on how to control soil erosion have led to much insight on erosion as a natural process. Medical research has greatly improved our general understanding of animal physiology. Some basic research, on the other hand, is quickly applied. X rays were discovered by German physicist Wilhelm Roentgen as the result of basic research, as mentioned by Sagan in the previous quote. He published his findings in late December of 1895, and in early February of 1896, X rays were being used by medical doctors to study a fracture in the broken bone of one patient in the United States and to find a bullet in the leg of a gunshot victim in Canada.[17]

## WHY UNDERSTAND SCIENCE?

It should be obvious that individuals pursuing careers in science and in teaching science should understand the fundamental nature of the scientific endeavor, of which this book is an introduction. It is important, though, to realize that all educated individuals, not just scientists, can benefit from gaining this knowledge as part of a liberal arts education. Many careers, while not specifically scientific, involve the use of scientific information. Business people frequently use the results of marketing surveys to make important decisions. Without understanding the role of survey design and methods of analysis on the results obtained, expensive mistakes can be made. Experience in experimental design and analysis, even in topics unrelated to market surveys, can provide insight into the strengths and limitations of the process. Similarly, so-called "scientific" approaches to analyzing the stock market require careful scrutiny before one relies on such techniques for investing money.[18] Experience in scientific research, if nothing else, provides insight on what questions to ask about topics related to science.

Citizens in a democracy have an obligation to be informed on the issues faced by their governments. Environmental policy, social welfare programs, military planning, and economic policy, for example, have important scientific components that often get overwhelmed by the emotional aspects of the problems. Scientifically literate citizens should demand that scientific research be used properly in the decision-making process. This means that, when possible, appropriate research be conducted and carefully scrutinized before an important decision is made. Scientifically literate citizens also should demand that scientific conclusions used in making policy decisions are chosen because the conclusions are the results of well-done science and not because they support a predetermined political position.

Individuals should understand science in order to make important decisions about themselves and their families. Medical research reported in the press often contradicts previous medical practice. Should an individual change diet or exercise patterns based on the latest research without first consulting a physician? Such decisions require an understanding of the inherent unreliability of "frontier" research, as discussed in Chapter

2. Intelligent consumer choices require careful comparisons of competing products. If one relies on analyses of products reported in magazines or books before buying cars or computer equipment or underwear, the analyses should be scrutinized before the conclusions can be trusted. Changing approaches to the education of children should be done because there is strong evidence that the changes will improve education, not because they merely sound good or fit into the latest educational fad. Parents have a responsibility to their children to see that there is a solid basis for changing educational policies. Scientifically literate parents also should be able to determine if the evidence used to support such changes is appropriate to the problem. Experience in science makes such scrutiny easier and more meaningful.[19]

Finally, the public is bombarded by *pseudosciences*, or phenomena improperly explained as scientific. Creation "science," psychic phenomena, some medical fads, UFOs, astrology, and many other pseudoscientific topics range from silly to dangerous to possibly legitimate areas of knowledge. Separating sense from nonsense on such matters requires an understanding of science in order to realistically evaluate the claims. Such evaluations require a thoughtful skepticism which comes from scientific thinking.

The next chapter introduces some philosophical aspects of science, most notably, the scientific method as a way of understanding things. This leads to Chapter 3, which gives an overview of the application of the scientific method in the process of research. Scientific knowledge must be shared with and evaluated by other experts, which is the topic of Chapter 4. The ethical conduct of scientists is the subject of Chapter 5. Science requires the use of critical thinking skills, which are discussed in Chapter 6, and pseudoscience is the topic of chapters 7 and 8. Though far from being a thorough discussion of science, this book is intended to introduce some of the more important aspects of the topic. The concluding chapter is an overview of how scientific knowledge is acquired.

This chapter concludes with a quote from the British scientist Thomas Henry Huxley. He was writing in 1868 about the need for a liberal education (liberal in the sense of "liberating"), and while his comments are not limited only to science, they clearly drive home the value of advancing scientific knowledge and science education.

---

Suppose it were perfectly certain that the life and fortune of every one of us would, one day or other, depend on his winning or losing a game at chess. Don't you think that we should all consider it to be a primary duty to learn at least the names and the moves of the pieces; to have a notion of a gambit, and a keen eye for all the means of giving and getting out of check? Do you not think that we should look with disapprobation amounting to scorn, upon the father who allowed his son, or the state which allowed its members, to grow up without knowing a pawn from a knight?

Yet it is a very plain and elementary truth, that the life, the fortune, and the happiness of every one of us, and, more or less, of those who are connected with us, do depend on knowing something of the rules of a game infinitely more difficult and complicated than chess. It is a game which has been played for untold ages, every man and woman of us being one of the two players in a game of his or her own. The chess-board is the world, the pieces are the phenomena of the universe, the rules of the game are what we call the laws of Nature. The player on the other side is hidden from us. We know that his play is always fair, just and patient. But we also know, to our cost, that he never overlooks a mistake, or makes the smallest allowance for ignorance. To the man who plays well, the highest stakes are paid, with that sort of overflowing generosity with which the strong shows delight in strength. And one who plays ill is checkmated—without haste, but without remorse.

Well, what I mean by Education is learning the rules of this mighty game. In other words, education is the instruction of the intellect in the laws of Nature, under which name I include not merely things and their forces, but men and their ways; and the fashioning of the affections and of the will into an earnest and loving desire to be in harmony with those laws. For me, education means neither more nor less than this.

Thomas Henry Huxley, 1868[20]

---

## PROBLEMS

1. What distinguishes science from other approaches to gaining knowledge?

2. Give an example of a research topic that requires an interdisciplinary approach. Explain.

3. Think of a technology you use regularly. Speculate on what basic research and applied research likely went into the development of that technology.

4. Explain to a person who has never really thought about the value of education why learning about science is important.

## NOTES

1. Rutherford, F. James, and Ahlgren, Andrew, 1990, *Science for All Americans*, New York: Oxford University Press, p. vi.

2. Strahler, Arthur N., 1992, *Understanding Science: An Introduction to Concepts and Issues*, Buffalo: Prometheus Books, p. 8.

3. Bunge, Mario, 1984, "What Is Pseudoscience?" *The Skeptical Inquirer* 9: 36–46; Strahler, 1992, p. 223–69.

4. Strahler, 1992, p. 11.

5. Frenzel, Gottfried, 1985, "The Restoration of Medieval Stained Glass," *Scientific American* 252, no. 5: 126–35.

6. See, for example, Knudsen, Vern O., 1963, "Architectural Acoustics," *Scientific American* 209, no. 5: 78–92.

7. Damon, P. E.; Donahue, D. J.; Gore, B. H.; Hatheway, A. L.; Jull, A. J. T.; Linick, T. W.; Sercel, P. J.; Toolin, L. J.; Bronk, C. R.; Hall, E. T.; Hedges, R. E. M.; Housley, R.; Law, I. A.; Perry, C.; Bonani, G.; Trumbore, S.; Woelfi, W.; Ambers, J. C.; Bowman, S. G. E.; Leese, M. N.; and Tite, M. S.; 1989, "Radiocarbon Dating of the Shroud of Turin," *Nature* 337: 611–15.

8. Kline, Morris, 1955, "Projective Geometry," *Scientific American* 192: 80–86.

9. Peck, Robert McCracken, 1988, "A. H. Thayer: Art of the Invisible," *International Wildlife* 18, no. 5: 38–43.

10. See Root-Bernstein, Robert, 1997, "Art, Imagination and the Scientist," *American Scientist* 85: 6–9. See also Root-Bernstein, Robert S.; Bernstein, Maurine; and Garnier, Helen, 1995, "Correlations Between Avocations, Scientific Style, Work Habits, and Professional Impact of Scientists," *Creativity Research Journal* 8, no. 2: 115–37.

11. See, for example, Neev, David, and Emery, K. O., 1995, *The Destruction of Sodom, Gomorrah, and Jericho: Geological, Climatological, and Archaeological Background*, New York: Oxford University Press. That information useful to scientists is contained in the

Bible is not to suggest that the Bible is a purely historical work, a topic discussed in Chapter 8.

12. Strahler, 1992, p. 13.

13. Mackinder, H. J., 1887, "On the Scope and Methods of Geography," *Proceedings of the Royal Geographical Society and Monthly Record of Geography* 9, no. 3: 141–61. Quote is on p. 154.

14. Sagan, Carl, 1995, *The Demon-Haunted World: Science as a Candle in the Dark*, New York: Random House, p. 397.

15. Williams, L. Pearce, 1965, *Michael Faraday*, New York: Basic Books, Inc., p. 196.

16. Sagan, 1995, p. 398.

17. Eisenberg, Ronald L., 1992, *Radiology: An Illustrated History*, St. Louis: Moseby Year Book, p. 22–28, 63–64.

18. For a discussion of real scientific approaches to stock market analyses, see Taubes, Gary, 1998, "Wall Street Smarts," *Discover* 19, no. 10 (October): 105–13.

19. For a related and amusing (and frightening) story on textbook selection for school children, see Feynman, Richard P., 1985, *Surely You're Joking, Mr. Feynman!: Adventures of a Curious Character*, New York: Bantam Books, p. 262–76.

20. Huxley, Thomas Henry, 1868, "A Liberal Education; and Where to Find It," delivered to the South London Working Men's College, in *The Major Prose of Thomas Henry Huxley*, ed. Alan P. Barr, Athens, Georgia: University of Georgia Press, 1997, p. 205–23. Quote is on p. 208–9.

# CHAPTER 2

# THE PHILOSOPHY OF SCIENCE

*First thesis:* We know a great deal. And we know not only many details of doubtful intellectual interest but also things which are of considerable practical significance and, what is even more important, which provide us with deep theoretical insight, and with a surprising understanding of the world.

*Second thesis:* Our ignorance is sobering and boundless. Indeed, it is precisely the staggering progress of the natural sciences . . . which constantly opens our eyes anew to our ignorance. . . . With each step forward, with each problem which we solve, we not only discover new and unsolved problems, but we also discover that where we believed that we were standing on firm and safe ground, all things are, in truth, insecure and in a state of flux.

*Karl Popper, 1985*[1]

This chapter continues the discussion of science begun in Chapter 1 by introducing some concepts important in understanding scientific knowledge. Much of this material fits into the category of the *philosophy of science*, which helps us determine how we gain an understanding of scientific phenomena (related to Popper's first thesis) and also understand the limitations and tentative nature of our knowledge (see Popper's second thesis). The discussion will begin with an introduction to the seemingly simple concept of scientific statements.

## SCIENTIFIC STATEMENTS

*Scientific statements* are attempts to describe or explain real phenomena. Examples are "force equals mass times acceleration," "Americans tend to vote in an attempt to improve their personal economic situation," and "antibiotics are effective at fighting infections." All scientific statements are not equally useful, though; some are more reliable and some are more specific than others.

An important concept in science is that there are no absolute scientific "truths" that can be used. Scientists must approach their work with the understanding that all scientific statements may be wrong and that probability is important. The scientist must appreciate the idea that some knowledge is less likely to be (or has a lower *probability* of being) wrong than other knowledge. Most scientists consider the likelihood that the sun will not rise tomorrow to be extremely small and can rely on sunrise every day in scientific reasoning. Keep in mind, though, that it is not good scientific practice to say there is absolutely no possibility that the sun will not rise tomorrow. On the other hand, if the weather has been hot and dry for the past week, the scientist should not automatically assume that the weather will be hot and dry next week, because the probability that the weather will change is much greater than the probability that the sun will not rise. Scientists must consider the reliability of scientific statements.

For a statement to be scientific, it must be *testable.* "Alligators eat more food than is eaten by crocodiles of the same size" is a scientific statement because the food intake of each can be measured and compared. "Alligators are prettier than crocodiles" is not scientific because you cannot measure "pretty." Pretty is an aesthetic value judgment, and as such it is beyond the scope of science.

Scientific statements must not only be testable, but the tests must be *reproducible.* Experiments must be presented such that someone else can do the same experiment. If others cannot get the same result, then the original results cannot be considered scientific. For example, a scientist measures the food intake of alligators and crocodiles and presents the results as a general statement. A second scientist comes along later and duplicates the first experiment and finds a different relationship. Assuming the measurements were made and analyzed in the same way, the first conclusion has to be considered applicable only to the time and place it was made, and more studies should be done to find a more reliable statement as to the relative amounts of food eaten by alligators and crocodiles.

Some areas of science, like chemistry, are more experimental in nature than others, like paleontology. Directly reproducing experiments in chemistry is easier than reproducing the findings resulting from the analysis of fossils. The philosophy of science, as described in this chapter, is more applicable to the experimental sciences than those in which experiments play a less important role.

*Scientific knowledge* should represent the most accurate view of the real world that is possible. It involves the most complete description of the phenomena we study and the most accurate explanation of why things are the way they are, given our current abilities to investigate.[2] Keep in mind, though, that all scientific knowledge may be found to be imperfect, and some currently accepted knowledge may be found to be completely wrong. Science is an ongoing process; there is no end product.

The way science is done is called *the scientific method*, though in fact many methods exist. The scientific method is the way in which scientific knowledge is gained. It has evolved into a system in which the likelihood of being wrong is reduced by methods used by individual scientists and by the skepticism and interactions of the *community of scientists* working together. It is the duty of all scientists to see that science is done properly, as discussed in chapters 4 and 5.

## SCIENTIFIC METHODS

Ancient Greeks, like Aristotle, used reason to explain things, sometimes referred to as the *classical approach* to science. They felt that if the argument is sound, then the conclusion is trustworthy. This method still works well in much of mathematics and logic, but in science it is not enough, because when real things are being studied, what appears correct in the mind and what exists in nature are often not the same thing. Bertrand Russell nicely summed up the problem of only using logic in science: "Aristotle could have avoided the mistake of thinking that women have fewer teeth than men, by the simple device of asking Mrs. Aristotle to keep her mouth open while he counted. He did not do so because he thought he knew. Thinking that you know when in fact you don't is a fatal mistake, to which we are all prone."[3] Scientists developed the empirical approach to check our thoughts with reality. Of course, Aristotle is a towering figure in both philosophy and science, and just because his scientific method was modified does not diminish the significance of his contributions: Logic remains a fundamental tool of science, just not the only one.

*Empiricism* arrived in Western culture at the time of the Renaissance. Empiricists feel that reason alone is not sufficient—ideas must be tested in the real world, thus making observation, measurement, and experiment central aspects of science.

---

[P]ure logical thinking cannot yield us any knowledge of the empirical world; all knowledge of reality starts from experience and ends with it. Propositions arrived at by purely logical means are completely empty as regards reality. Because Galileo saw this, and particularly because he drummed it into the scientific world, he is the father . . . of modern science. . . .

Albert Einstein, 1954[4]

---

Empiricism can be approached in two ways: induction and deduction. In *induction*, the scientist goes out and measures aspects of the phenomenon under study. Then those measurements are analyzed and generalizations, also called *theories*, are made from the analysis. Charles Darwin caught finches on the Galapagos Islands. He later noticed that the shapes of the birds' beaks were different on each island. After studying the beaks, he concluded that each shape seemed to serve a purpose suited to the conditions on a particular island. For example, finches have short, fat beaks on islands where the main food source is nuts, which have to be broken; on islands where the main food source is insects found inside holes in trees, the finches have long, skinny beaks. Darwin developed a theory that at some time in the past, one type of finch arrived at the islands and then evolved differently on each island. He later extended this theory of evolution to all life forms;[7] thus, the most important theory in the biological sciences was developed using inductive reasoning: *data collection–analysis–theory*.

In the twentieth century, an alternative to induction has gained popularity. This is *deduction*. In deduction, you start with a theory (often simply an educated guess at this point), then move on to data collection and then analysis *(theory–data collection–analysis)*. A problem with induction is that there are few guidelines to tell the scientist which measurements to make and how to make them. If you start with a theory, then you have a better idea of what information is needed to test it.

---

In order to act in a scientific manner ... two things are necessary: fact and thought. Science does not consist only of finding the facts; nor is it enough only to think, however rationally. The processes of science are characteristic of human action in that they move by the union of empirical fact and rational thought, in a way which cannot be disentangled. There is in science, as in all our lives, a continuous to and fro of factual discovery, then of thought about the implications of what we have discovered, and so back to the facts for testing and discovery—a step by step of experiment and theory, left, right, left, right, for ever.

J. Bronowski, 1978[5]

---

Deduction is believed to be more efficient than induction and less likely to lead scientists astray. When little is known about a subject, however, scientists may find it advantageous to use induction in order to gain an initial understanding of the phenomenon. Such was the case of Darwin and the finches and is now the case with some scientists studying the surfaces of planets with recently acquired images via spacecraft.

---

I often say that when you can measure what you are speaking about, and express it in numbers, you know something about it; but when you cannot measure it, when you cannot express it in numbers, your knowledge is of a meagre and unsatisfactory kind: it may be the beginning of knowledge, but you have scarcely, in your thoughts, advanced to the stage of science....

William Thompson (aka Lord Kelvin), 1883[6]

---

The best-known scientific method is *positivism*. Positivism starts with a scientist thinking about the phenomenon he or she plans to study. From this image of the real world, the scientist states a *hypothesis*—a possible explanation of how something works. Hypotheses are based on what the scientist thinks is the likely situation and may come from past observations by the scientist, the findings of other scientists, or simply hunches.

The word *theory* is essentially synonymous with hypothesis, although many people use hypothesis for a statement before it is tested and theory when it has passed a test. John Ziman further refines the concept of theory by stating that theories should meet three criteria. First, a theory should be *rational*, or logically consistent. "Any ambiguity, imprecision, or inconsistency in the chain of argument leading from the hypothesis to its implications can give rise to serious errors or uncertainties in the validation process."[8] Second, a theory should be *relevant*. It should be clear how the theory relates to the real world. Third, a theory should be *extensible*, or able to explain more than was originally intended. As discussed previously, Darwin put forth a theory to explain the variation in characteristics of finches on the Galapagos Islands but then realized that the theory applied to all life forms.[9]

In positivism, hypotheses are tested to find out if they are likely to be correct. This is generally done by experiment—carefully collecting information and analyzing it. Experiments are sometimes done in the laboratory, sometimes in the field (outside the laboratory), and sometimes in mathematical simulations. If the hypothesis fails the test, then the scientist needs to revise it and test the new version. If the hypothesis passes the test, then the scientist can test it in other situations or refine it and test the new version. As scientists keep revising and testing hypotheses, the hypotheses should become more accurate descriptions of reality.

A modification of the positivist approach is *falsification*, which was introduced by Karl Popper.[10] No matter how many times a theory passes a test, scientists can never be sure that it will pass all future tests. The falsification approach gets around this problem by having the scientist make a hypothesis and then try to show that it is *not* correct. If a scientist can find one case where the hypothesis legitimately fails the test, the hypothesis can be rejected, whereas positivists cannot truly accept a theory until it passes all possible tests. A commonly used hypothetical example of the value of falsification is that seeing one million swans that are white does not prove that "all swans are white," but finding one that is black disproves the theory. Once a hypothesis is falsified, it is modified and experiments are designed to try to falsify the new version. When a theory has survived many vigorous attempts to falsify it, it is considered reliable.

In both positivism and falsification, hypotheses are tested by experiment. If the hypothesis is supported by such tests, then scientists have more confidence in it. When a hypothesis is not verified by an experiment, then scientists have to decide if the problem arises from inadequacies of the hypothesis or from problems with the experiment. Isaac Newton's mechanics, for example, were found to predict accurately the orbits of most of the planets, with the exceptions of Uranus and Mercury. In studying the orbit of Uranus, scientists decided that the theory may be correct, and the problem in predicting the orbit may have been with the experiment. Using the principles of Newtonian mechanics, it was suggested that an unknown planet existed that affected the orbit. Eventually, the planet Neptune was discovered, which accounted for the discrepancies in the predicted and observed orbit of Uranus. No new planet has been discovered to account for the discrepancy between the actual and predicted orbit of Mercury, where the Newtonian theory has been replaced by Albert Einstein's theory of relativity to accurately predict the orbit of Mercury. In the test of Newtonian mechanics involving the orbit of Uranus, the theory was found to be correct and the experiment needed revision. In the test using the orbit of Mercury, the experiment was correct and the theory needed changing. A difficult and crucial part of science is determining which action is required.[11] According to Ziman, "It is part of the art of scientific investigation not to be daunted by a few negative experimental results, and not to reject hypotheses that seem to have been 'falsified' by a few disconfirmatory instances but which still have a great deal going for them in other ways."[12]

Through repeated testing of a hypothesis, scientists gain greater confidence that it reflects reality reasonably well. Keep in mind, though, that having more confidence in a hypothesis is fundamentally different from believing it to be true. If most scientists in a particular discipline are convinced that a hypothesis will pass every conceivable test, then it is called a *law*. There are no rules for when a theory becomes a law—just that almost all scientists familiar with it have to agree. More specifically, scientific laws are general statements that apply over wide ranges of time and space, are found to be correct in every situation, and have an extremely small likelihood of being found incorrect.

In reality, scientists never know if a theory is going to be true in all cases because they cannot do all possible tests. Scientists simply gain more confidence in the theory as it continues to pass tests. Some of Newton's "laws" of motion were shown to be incorrect in some applications when Einstein applied them to predict the motion of astronomical bodies. Newton's "laws" have been modified by Einstein's theory of relativity under such circumstances, as discussed in predicting the orbit of Mercury. Newton's "laws" are not universally true, as technically laws should be, though they are highly reliable for use on Earth and much of the solar system.

---

Few claims in science ... are accepted as final. But as evidence piles upon evidence and theories interlock more firmly, certain bodies of knowledge do gain universal acceptance.... [T]hey ascend a scale of credibility from "interesting" to "suggestive" to "persuasive" and finally "compelling." And given enough time thereafter, "obvious."

Edward O. Wilson, 1998[13]

---

Laws are quite rare in science. There are no real laws in most scientific disciplines; the subject matters are simply too complex for scientists to have found statements that fit the criteria for laws. Laws are found mainly in physics and chemistry, though they are used in the other sciences. The ideal gas law in chemistry, for example, relates the volume, pressure, and temperature of gases. It is used also in atmospheric studies to understand the change of air temperature with altitude.

Because all scientific statements must be considered at best only tentatively correct, the phrase "scientific proof" should be used with extreme caution in scientific discussions. Strictly speaking, nothing in science is truly proved.

As an aside, the reliability of the scientific understanding of a phenomenon depends to some extent on the amount of research that has been done on the topic. Scientific knowledge can be divided into two categories: *textbook* and *frontier science*. Most of the scientific information found in textbooks is well-established knowledge and is considered highly reliable by most scientists. On the frontier, or "cutting edge," of research, though, hypotheses have not been tested sufficiently to warrant the trust put in textbook science. Most information found in textbooks has been tested many times, found to be reasonably accurate, and is not expected to change significantly, at least not in the near future. After much testing and debate, a small fraction of frontier science makes it into the textbooks, and the rest is deemed unreliable. Ziman makes this point by suggesting that "[t]he physics of undergraduate text-books is 90% true; the contents of primary research journals of physics is 90% false."[14] While scientists may report that new experiments have shown that a certain medication seems to increase hair growth on bald men, the press may report their study as proof of a cure for baldness. Quite often the press and the public fail to see the distinction between textbook (a cure has been found!) and frontier science (evidence that suggests that progress may be made . . . ).[15]

### RECENT DEVELOPMENTS
### IN THE PHILOSOPHY OF SCIENCE

The description of positivism and falsification previously presented suggests that scientific knowledge about the real world is gained through experiments, where observations determine the validity of theories, and that there is a relatively steady increase in our understanding of scientific phenomena. Starting in the 1950s, revisions to these concepts have been introduced and debated.

It is often assumed that science is *objective*, in that evidence from the real world determines if a theory is correct, and the prejudices of the scientist do not affect the results presented. However, scientists are people, and they have their biases like everyone else, which adds a *subjective* element to science. As an extreme example, early in the twentieth century, racism was justified by some white scientists because that was their view of the world. They designed tests, mostly subconsciously, to confirm their preconceived ideas that males of European ancestry were intellectually superior to males of African ancestry and ignored any evidence that contradicted their theories.[16] More subtly, many scientists who trust a theory for many years are unlikely to discard the theory even if there is strong evidence against it.

One subjective element of science involves scientists' perceptions of the real world. We often assume that what we see is what is really there, but psychologists point out that what we see is influenced by what we expect to be there, and what we expect is influenced by past experience. Scientists testing a theory, which is a product of past experience, allow the theory to guide them in interpreting their experimental results. As Harold Brown states, "The scientist does not record everything he observes but only those things which the theories he accepts indicate are significant."[17] Science is said to be *theory-laden*, in that theories strongly influence how science is done. Much scientific knowledge is gained by investigating anomalous phenomena which do not fit with existing theory, like the search for the causes of the unpredictable orbits of Uranus and Mercury already mentioned. Brown also states that ". . . we should note that in the important case in which a scientist identifies a phenomenon as anomalous or problematic he is clearly observing its meaning in terms of the theories he holds, for if he had no beliefs about what ought to occur in the situation in question, no occurrence could be perceived as problematic."[18] Therefore, a theory tells scientists what to look for in interpreting the test of the theory, which, of course, is an imperfect way to test a theory.

Although truly objective science is impossible. many scientists consider objectivity to be a goal worth pursuing. It is the duty of scientists to minimize subjectivity in their own work and to point out subjectivity in the works of others and therefore to allow science to progress toward a more objective description of the real world and how it works. As Milton Rothman states, "[o]bjectivity in science is obtained not by the individual scientist, who is always subjective, but by the scientific community as a whole. The examination of a scientific claim by peers with other points of view is the best way to grind rough edges off a theory and to detect any errors that may exist."[19] Such edge grinding is discussed in Chapter 4.

Another relatively recent development in the philosophy of science is the idea that whereas science deals with empirical, or *observable*, phenomena, theories usually contain *unobservable* parts. For example, if you go to the doctor with a fever, the doctor may hypothesize that you have a viral infection. However, tests for viruses in the body are not readily available to most doctors. Various tests can be done, which may reduce the likelihood that the fever is caused by bacterial infection or other causes. Without ever seeing or otherwise detecting an actual virus, the doctor may conclude that the illness is viral and suggest appropriate treatment. Scientists, like physicians, need to find ways to connect the unobservable parts of a theory with real phenomena. Economists cannot directly determine the overall state of a national economy, so they rely on indicators such as stock market trends, unemployment rates, and others. Paleontologists cannot observe dinosaur behavior, so they must rely on indirect evidence from fossilized bones and other information. Of course, scientists must keep in mind how the use of such indicators affects the reliability of their conclusions.[20]

The discussion of science thus far has given the idea that science progresses slowly and steadily through research that builds on previous knowledge. Such is not always the case. Thomas Kuhn studied science from a historical perspective and concluded that wholesale changes in thinking sometimes occur. Most of the time, the scientists in a discipline operate under a *paradigm*, or general view of the subject matter. The paradigm provides a framework for a discipline in terms of what to study and how to study it. For example, in the second century A. D., Ptolemy suggested that the Sun revolves around Earth, and for centuries astronomers conducted research with that as a basic assumption. In the sixteenth century, however, Copernicus found convincing evidence that Earth revolves around the Sun. Eventually, most astronomers were convinced that the Ptolomaic paradigm was incorrect, and it was replaced by the Copernican paradigm. The transition from one paradigm to another is called a *scientific revolution*, and it results in an often drastic transformation in the way scientists think about their discipline. After a revolution, a new paradigm guides thinking about the topics studied. Disciplines may operate within a paradigm for decades or even centuries before a revolution takes place, and there is debate as to whether all disciplines have revolutions. Some disciplines, like sociology, deal with such complex topics and unpredictable phenomena that they do not have paradigms at all; they may have several competing views of their subject matter.[21]

In the philosophy of science today, three perspectives provide modifications to the positivist and falsificationist views: postpositivist empiricism, relativism, and realism.[22] Positivism and falsification, as previously described, work on the assumption that science progresses toward a true description of observable aspects of reality. *Postpositivist empiricism* emphasizes that the unobservable parts of theories are important but are not directly verifiable by experiment. Therefore, the observable aspects of scientific knowledge, after they have been tested, are more trustworthy than the unobservable parts, which can be tested only indirectly. While scientists may never know if a theory is true, they can tell if it passes experimental tests reasonably well or helps to solve problems. We may never know if our current understanding of mechanics is an absolutely true explanation of part of reality, but if it is used to design bridges and they do not fall down, then it is a theory worth trusting.

In *relativism*, the emphasis is on the role of social factors in the conduct of science. Relativists argue that because of the theory-laden nature of science, the scientific method is viciously circular, as has been discussed. Therefore, scientific research does not proceed toward objective truth. Instead, it affects the paradigm, or perspective on reality shared by the members of a discipline. In its moderate form, relativism suggests that science is a problem-oriented endeavor that can effectively solve problems even if it does not explain objective reality. In its extreme version, relativism states that science is a purely social, or even individual, function with little or no relationship to the real world.[23] Few scientists believe the latter version of relativism.

*Realism* is based on the idea that there is a real world which is separate from our perceptions of it and that the methods of science allow us to get closer to understanding that reality, even the unobservable parts. According to realists, while science may never attain absolute truth, it can get close to it. If theories are approximately true, they will describe reality reasonably well. Therefore, the theory-laden nature of science, emphasized by postpositivist empiricists, is not a serious defect. According to realists, there is ample evidence that as a means of understanding the world, science works. Therefore, scientists must be describing the world reasonably accurately, including the unobservable parts of it. (The book you are reading was written largely from a realist perspective.)

It is not clear if one of the three alternative approaches to science will win out and become the dominant paradigm in the philosophy of science, as positivism (along with falsification) was earlier in the twentieth century.

The reader should understand that the philosophy of science is a complex subject, and this chapter is a highly simplified overview of it. Many of the individual sentences contained here come from countless hours of contemplation and discussion and numerous books and articles by philosophers and scientists. Also, many aspects of the philosophy of science are not mentioned at all.

## PROBLEMS

1.  How does probability relate to scientific statements?

2.  Why should scientific tests be reproducible?

3.  Why was empiricism an improvement over the classical approach to science?

4.  Define and distinguish between

    A. hypothesis, theory, and law

    B. induction and deduction

    C. positivism and falsification

    D. textbook and frontier science

    E. objectivity and subjectivity

5.  What is meant by the phrase "science is theory-laden"?

6.  "Science can never be truly objective." What are your thoughts on this statement?

NOTES

1. Popper, Karl R., 1985, in Miller, David, ed., *Popper Selections,* Princeton: Princeton University Press, p. 7.

2. Strahler, Arthur N., 1992, *Understanding Science: An Introduction to Concepts and Issues,* Buffalo: Prometheus Books, p. 8.

3. Russell, Bertrand, 1962, *Essays in Skepticism,* New York: Philosophical Library, Inc., p. 70.

4. Einstein, Albert, 1954, *Ideas and Opinions,* New York: Bonanza Books, p. 271. (From a lecture titled "On the Method of Theoretical Physics" delivered in 1933.)

5. Bronowski, J., 1978, *The Common Sense of Science,* Cambridge: Harvard University Press, p. 30.

6. Thompson, William, 1884, "Electrical Units of Measurement," in *The Practical Applications of Electricity. A Series of Lectures Delivered at The Institution of Civil Engineers, Session 1882–83,* London: The Institution of Civil Engineers, p. 149–76; quote is on p. 149.

7. The development of Darwin's Theory was much more complex than presented here; the reader is encouraged to read a more complete discussion, as can be found in Mayr, Ernst, 1982, *The Growth of Biological Thought: Diversity, Evolution, and Inheritance,* Cambridge: The Belknap Press of Harvard University Press, p. 394–534.

8. Ziman, John, 1984, *An Introduction to Science Studies: The Philosophical and Social Aspects of Science and Technology,* Cambridge: Cambridge University Press, p. 46.

9. Ibid., p. 28–29.

10. See Popper, Karl R., 1959, *The Logic of Scientific Discovery,* New York: Basic Books. (Originally published as *Logik der Forschung* in 1934.) See also Popper, Karl R., 1963, *Conjectures and Refutations: The Growth of Scientific Knowledge,* 2d ed., New York: Basic Books.

11. Brown, Harold I., 1977, *Perception, Theory and Commitment: The New Philosophy of Science,* Chicago: Precedent Publishing Inc., p. 97.

12. Ziman, 1984, p. 47.

13. Wilson, Edward O., 1998, *Consilience,* New York: Alfred A. Knopf, p. 59.

14. Ziman, John, 1978, *Reliable Knowledge: An Exploration of the Grounds for Belief in Science,* Cambridge: Cambridge University Press, p. 40.

15. Bauer, Henry, H., 1992, *Scientific Literacy and the Myth of the Scientific Method,* Urbana: University of Illinois Press, p. 32, 103–4.

16. Gould, Stephen Jay, 1981, *The Mismeasure of Man,* New York: W. W. Norton & Co.

17. Brown, 1977, p. 90.

18. Ibid.

19. Rothman, Milton A., 1991, *The Science Gap: Dispelling the Myths and Understanding the Reality of Science,* Buffalo: Prometheus Books, p. 177.

20. Bunge, Mario, 1983, *Understanding the World,* Vol. 6, *Treatise on Basic Philosophy,* part 2, Dordrecht: D. Reidel Publishing Co., p. 85–91.

21. Kuhn, Thomas S., 1970, *The Structure of Scientific Revolutions,* 2d ed., Chicago: University of Chicago Press.

22. Rhoads, Bruce L., and Thorn, Colin E., 1994, "Contemporary Philosophical Perspectives on Physical Geography with Emphasis on Geomorphology," *Geographical Review* 84: 90–101.

23. See Schick, Theodore, Jr., and Vaughn, Lewis, 1995, *How to Think about Weird Things: Critical Thinking for a New Age*, Mountain View, California: Mayfield Publishing Co., p. 69–95.

# CHAPTER 3

# RESEARCH

The Blind Men and the Elephant
by John Godfrey Saxe[1]

### I.

It was six men of Indostan
  To learning much inclined,
Who went to see the elephant
  (Though all of them were blind),
That each by observation
  Might satisfy his mind.

### II.

The *First* approached the elephant,
  And happening to fall
Against his broad and sturdy side,
  At once began to bawl:
"Bless me!—but the Elephant
  Is very like a wall!"

### III.

The *Second*, feeling of the tusk,
  Cried, "Ho!—what have we here
So very round and smooth and sharp?
  To me 't is mighty clear
This wonder of an Elephant
  Is very like a spear!"

### IV.

The *Third* approached the animal,
  And, happening to take
The squirming trunk within his hands,
  Thus boldly up and spake:
"I see," quoth he, "the Elephant
  Is very like a snake!"

### V.

The *Fourth* reached out his eager hand,
  And felt about the knee.
"What most this wonderous beast is like
  Is mighty plain," quoth he;
"'T is clear enough the Elephant
  Is very like a tree!"

### VI.

The *Fifth*, who chanced to touch the ear,
  Said, "E'en the blindest man
Can tell what this resembles most;
  Deny the fact who can,
This marvel of an Elephant
  Is very like a fan!"

### VII.

The *Sixth* no sooner had begun
  About the beast to grope,
Than, seizing on the swinging tail
  That fell within his scope,
"I see," quoth he, "the Elephant
  Is very like a rope!"

### VIII.

And so these men of Indostan
  Disputed loud and long,
Each in his own opinion
  Exceeding stiff and strong,
Though each was partly in the right,
  And all were in the wrong!

### Moral.

So, oft in theologic wars
  The disputants, I ween,
Rail on in utter ignorance
  Of what each other mean,
*And prate about an elephant*
  *Not one of them has seen!*

## INTRODUCTION

The men in Saxe's poem, representing people everywhere, were confronted with what is, in essence, a scientific problem. Elephants are part of the real world, and the blind men sought to understand what elephants are like. Their approach to solving the problem, however, was not scientific, and the resulting disagreement is understandable in light of the method used. Each collected a small amount of information, interpreted it, and then declared an answer to the problem. In modern science—that developed in the past 400 years or so—ways have been found to test initial ideas to see if they have merit before conclusions are drawn, and the conclusions are qualified because it is clear that they may be wrong. By using scientific research, people have learned about highly complicated and elusive topics. Most research done today is on phenomena far more difficult to study than determining what an elephant looks like, even without the benefit of sight.

This chapter provides a selective overview of scientific research. It is not a "how to" manual for doing research, it just introduces some aspects of the process. Scientific research is done in many different ways, from finding fossils and interpreting them to laboratory studies of subatomic phenomena to mathematically simulating the atmosphere in a computer model to studying trends in historical census data, to name a few. Because of this diversity, no chapter or even whole book can provide a complete view of research. Some of the common threads in all scientific research, however, will be introduced here.

## SELECTING A TOPIC TO STUDY

Choosing a scientific problem to research involves many factors. Imagine that scientific knowledge is a map of the real world. The map is highly reliable in some sections (e.g., laws), but most of it is at varying degrees of trustworthiness. Some parts within the map are still uncharted territory, and of course, the map needs to be extended beyond its present boundaries into *terra incognita*. A critical review of the literature in any field of science typically will show many of the imperfections and limitations of the "map," and these are especially apparent to the trained eye.[2] Most scientists have no problem identifying potential topics to study; the difficulty lies in choosing which of them to pursue.

The topic that a scientist or team of scientists chooses to study usually involves some combination of interest in the phenomenon, technical expertise and resources needed to do the study, and ability to get funding for the project. Research requires such a large amount of intellectual energy and time that a scientist should feel a deep interest in the topic in order to do a good job; a bored scientist is unlikely to make great discoveries. A given research project will require certain expertise, such as knowledge of statistical tests or field methods, and certain equipment to be completed. The scientist must consider whether she or he possesses the requisite skills and equipment or will be able to obtain them in order to do the study. While some research can be done with small amounts of money and readily available equipment, funding is often necessary to pay for the project. If the latter is the case, then a funding organization will have to be convinced that the project is worth paying for before it can be started.[3]

## HYPOTHESES

Once a general topic has been chosen, the next step is to generate a *hypothesis*. As discussed in the previous chapter, hypotheses are possible explanations of some phenomenon. An example of a hypothesis is Christopher Columbus's suggestion that one could travel from Europe to the Orient by sailing west. This was a good hypothesis in that it was clearly stated and certainly testable. Columbus's voyages served as experiments testing it, and while he thought he confirmed the hypothesis, in fact, the situation was far more perplexing than he anticipated; the Americas were an unexpected complicating factor.[4] Hypotheses vary tremendously in their precision and complexity, depending on such factors as how much is known about the topic and the style of the scientist. One hypothesis may merely state that "variable A has an effect on variable B," while another may present an elaborate equation precisely predicting the relationship between A and B. Hypotheses can be elaborately derived estimations of a phenomenon, or they may be simple hunches.

Science typically is presented as a purely logical enterprise, and ideally at least, much of science is just that. However, when it comes to the generation of hypotheses, logic is but one thought process used. Hypotheses may come from careful and logical analyses of problems, but they may also come from the untamed reaches of the imagination. In this way, science is an art and as creative an endeavor as literature, painting, or music. Hypotheses can even arise from dreams, as was the case of the German chemist F. A. Kekulé. He tells of having a dream in which a snake is biting its own tail. He awoke suddenly from this dream with the kernel of a hypothesis that the chemical compound benzene is shaped like a ring and stayed up the rest of the night developing the idea. Confirmation of this hypothesis led to great advancements in organic chemistry.[5] Gerald Holton discussed the imagination of Albert Einstein, where a ". . . haunting picture at age sixteen of chasing or riding on a beam of light . . . contained the seed of his later work on special relativity."[6] When it comes to the development of scientific hypotheses, there are no rules which must be followed. Basically, if it works, do it. (Actually, there is one rule: you cannot steal someone else's idea. That is plagiarism and is one of the forms of scientific misconduct discussed in Chapter 5.)

A number of different thought processes are used in developing hypotheses. Some of the more common are analogy, induction, deduction, and intuition.[7] A hypothesis originating through *analogy*, simply put, is one where similar situations may lead to similar results. If we know that chimpanzees raise their young in certain ways, then we can hypothesize that gorillas do the same. A hypothesis generated by *induction* arises from observations of specific phenomena which are then hypothesized to reflect a general pattern. An ecologist may find that an insect species killed several trees and then hypothesize that that species is the cause of increased tree death throughout the forest. A hypothesis formulated through *deduction* is one based on better-established scientific knowledge. For example, an analysis of well-understood physical principles applied to the conditions of the atmosphere suggests that increasing the amount of carbon dioxide in the air will lead to an increase in global temperature. *Intuition* can be used to generate hypotheses, as well. Here, the scientist has little, if any, concrete information to guide him or her; the hypothesis is simply a statement as to what "seems right." Rarely is there no background information to guide the making of a hypothesis, but "gut feelings" often do play a role, as shown by the example of Kekulé and the snake dream discussed previously. While the modes of origin of many hypotheses can be grouped into these categories, one should not get the impression that a scientist sits down and says, "Today I think I'll use induction to generate a hypothesis." The creative process involved in hypothesis generation is complex and poorly understood.[8]

Chance sometimes plays a role in the development of hypotheses. In a *eureka intuition*, a scientist makes a sudden connection between a problem he or she has been thinking about and a common event. Scientists often immerse themselves in a problem, sometimes to the point of obsession, and through rather mysterious mental means, a simple observation may trigger the insight needed to develop a hypothesis. *Eureka* is Greek for "I have found it!" and the concept is generally associated with Archimedes. In the third century B.C., the king of Syracuse commissioned a crown made of pure gold but was later told that the craftsman had mixed silver with the gold. The king asked Archimedes to determine if he had been cheated, but without damaging the crown. Archimedes could not think of a way to test for silver until he took a bath and noticed that by lowering his body in the pool he displaced some of the water. In a sudden burst of inspiration, he realized that the volume of his body in the water is equal to the volume of water it displaces. He then knew that he could determine the volume of the crown by immersing it in water and then comparing the volume of the crown with the volume of a piece of pure gold of the same weight. Archimedes became so excited with his insight that he jumped out of the pool and ran home naked, shouting "Eureka!" His measurements then showed that the king, indeed, had been cheated, and the goldsmith was executed. The story of Isaac Newton's gaining insight on gravity after watching an apple fall out of a tree is another example of eureka intuition.[9]

In *serendipity*, an unexpected observation is made, and this, along with the wisdom of the observer, leads to a novel insight or important discovery. The term was coined by Horace Walpole in the 1750s and refers to a fairy tale about the "Three Princes of Serendip." (Serendip is now the country of Sri Lanka.) Walpole wrote to a friend that ". . . as their highnesses travelled, they were always making discoveries, by accidents and sagacity, of things they were not in quest of. . . ."[10] In the history of science, it is easy to find examples of important discoveries made by researchers with the wisdom and training to turn an unexpected observation into a valuable hypothesis. In the 1920s, Alexander Fleming was doing experiments on bacteria colonies and left some culture dishes on his workbench while he went on vacation. (Though a brilliant bacteriologist, he was not always tidy.) While talking to a colleague after returning to the laboratory, he picked up one of the dishes to demonstrate a point. In it he noticed that some of the colonies had become transparent because something had caused the bacteria to break down. Upon further study, penicillin was discovered. This antibiotic, of course, is one of the most important medical advances of the twentieth century. In the 1960s, A. A. Penzias and R. W. Wilson set out to measure the intensity of radio waves emitted by various parts of the galaxy. Unexpectedly, they detected microwaves not coming from any particular source. They initially assumed that they had malfunctioning equipment, but tests showed that the equipment was working properly. They eventually concluded that the mysterious microwaves, indeed, do exist and are background radiation from the "big bang" origin of the universe.[11]

---

It is of the utmost importance that the rôle of chance be clearly understood. The history of discovery shows that chance plays an important part, but on the other hand it plays only one part even in those discoveries attributed to it. For this reason it is a misleading half-truth to refer to unexpected discoveries as "chance discoveries" or "accidental discoveries." ...The truth of the matter lies in Pasteur's famous saying: "In the field of observation, chance favours only the prepared mind." It is the interpretation of the chance observation which counts. The rôle of chance is merely to provide the opportunity and the scientist has to recognize it and grasp it.

W. I. B. Beveridge, 1950[13]

---

While chance plays a role in science, it takes a good scientist to make the observation and then take advantage of it. Many thousands of people have seen apples fall off trees, but it took Newton to connect the action to gravity. Strange things happen in culture dishes all the time, but it took the scientific training and keen insight of Fleming and others to turn one of them into a lifesaving medicine. "A good scientist is discovery-prone," to quote Peter Medawar, meaning that they seek out and are always ready to learn from serendipitous opportunities.[12]

The ideas for hypotheses can come from almost anywhere, but there are a few general guidelines for evaluating them. First, a hypothesis should be logically consistent; one cannot test a hypothesis that contradicts itself. Second, it should be consistent with most of the available knowledge on the subject. Third, the hypothesis should offer an explanation of the phenomenon under consideration; seeking to explain things is the goal of science. Fourth, it should have the potential of adding to, or somehow being an improvement over, existing theory. Fifth, it must suggest consequences that are empirically testable; it should make a prediction that can be checked against reality. Prediction here refers not only to future events but also to what scientists will find if they look for what is suggested by the hypothesis. Thus, a hypothesis can predict that an asteroid will strike Jupiter next month or that one struck the earth 65 million years ago, or one can predict what type of substance will be created if certain chemicals are combined in a certain way.[14]

It is important to keep in mind that the true merit of a hypothesis can be determined only after it has been tested. No matter how good it looks in the mind of the scientist or how convincing it is to others, it is just an unchecked idea. Tests, usually in the form of experiments, are the next topic of this chapter.

---

Scientific hypotheses are not legitimate or illegitimate on account of their origin but on the strength of their tests ... : they are given test certificates rather than birth certificates.

Mario Bunge, 1967[15]

---

Finally, while many forms of thinking may be used in the generation of hypotheses, once the hypothesis has been formed, the rules of science change considerably. Testing of the ideas, analyzing the results of the tests, and evaluating the significance of the research requires the careful use of the mode of thought known as critical thinking.[16] Critical thinking is so important to science that it is treated in depth in chapters 6 and 8.

## EXPERIMENTAL DESIGN

The development of a hypothesis leads to the preparation of a plan to test it in such a way that reliable evidence is collected and analyzed. Since a major goal of science is to minimize the difference between our perception of reality and reality itself, the primary focus of experimental design is to make the test as objective as possible. A hypothesis is a scientist's best guess as to how something happens and is highly subjective. Therefore, testing that idea is supposed to maximize reality and minimize opinion.

The plan developed for an experiment should be detailed and comprehensive. All of the data should be collected in a consistent manner so that meaningful comparisons can be made. For example, if two researchers are involved in interviewing people in a psychological experiment, they should be given thorough guidelines as to how the questions are to be asked (one cannot be friendly while the other is nonemotional), how to respond if the person does not understand the question, etc.[17] Without a carefully prepared experimental plan, there is less control on how the data are collected, and the results are correspondingly less reliable. Plans commonly do get changed during an experiment, as will be discussed later in this chapter, but a comprehensive experimental design is necessary for useful and meaningful research.

One of the first decisions to be made in experimental design is what things are to be observed to shed light on the problem. Observable phenomena in science are those things that can be measured to get empirical information.[18] Often, direct observations are done in science, such as measuring the speed of snails in a study of their movement. When necessary, however, indirect observation is used. In such situations, indirect evidence is collected and inferences are made about the phenomena under study. One example of indirect observation is the use of fossils to study evolution. The fossils show the results of evolution but are not evolution itself. Another example is studies of the core of the Sun, which have been done by capturing neutrinos (subatomic particles given off during fusion) in deep mine shafts more than a kilometer below the surface of Earth. The neutrinos provide information about the processes acting in the sun's core, where they originate, but clearly, the scientists are not directly observing the core of the Sun.[19] Direct observation is desirable in science but often is not possible due to the nature of the subject being studied.

## *Variables*

The phenomenon under study in an experiment is called the *dependent variable*, while the factors that influence it are called *independent variables*. In most experiments, the most important independent variables are determined, and the relationships between them and the dependent variable are studied. Something that does not change throughout the experiment is called a *constant*. As a simple hypothetical example, an experiment is done to test the results of a model that predicts how long it will take for a ball to drop to the bottom of identical containers filled with various fluids. The density of the fluid is the independent variable, while the time of drop is the dependent variable, since the density determines the time. The force of gravity is important but does not change

---

I believe that every event in the world is connected to every other event. But you cannot carry on science on the supposition that you are going to be able to connect every event with every other event. . . . It is, therefore, an essential part of the methodology of science to divide the world for any experiment into what we regard as relevant and what we regard, for purposes of that experiment, as irrelevant. . . . We make a cut. We put the experiment, if you like, into a box. Now the moment we do that, we do violence to the connections in the world. . . . Therefore, when we practice science . . . , we are always decoding a part of nature which is not complete. We simply cannot get out of our own finiteness.

Jacob Brownowski, 1978[20]

---

and is, therefore, a constant in the experiment. (Of course, if the experiment were done on different planets, then gravity would be an independent variable instead of a constant.) There may be other influences on the drop time of the ball, such as how much the ball deviates from a perfect sphere and imperfections on the ball's surface, but the experimenter decides that they will have a sufficiently small effect on the results and, therefore, can be neglected. Determining which variables are to be included in an experiment is important because leaving out a significant one diminishes the value of the results while including insignificant ones increases the time, cost, and complexity of the work.

## Correlation and Causality

In this hypothetical ball drop experiment, identical balls can be dropped in identical containers, each filled with a fluid of a different density. The time it takes for each ball to drop is measured, and the results can be plotted on a graph with fluid density on one axis and time of drop on the other. The pattern of the dots on the graph suggests the relationship between the two variables. If the dots are aligned in an orderly fashion, such as a straight line or curve, then there is a strong relationship between the two. If there is no clear pattern to the dots, then the relationship is weak or nonexistent.

If the two variables appear to be related, however, that does not by itself suggest that the change in one necessarily causes the change in the other. *Cause and effect* is a complex topic in science, but simply put, one should not assume that there is a causal connection between two variables unless a well-founded theoretical explanation can be offered for the relationship. Accepted principles of physics indicate that fluid density determines the amount of resistance to falling that the ball experiences, so it can be said with confidence that fluid density causes the change in the speed of drop of the ball in the experiment. On the other hand, if the populations of rabbits and mice in an area follow the same pattern of increase and decrease over time, this correlation does not guarantee that the relationship between them reflects cause and effect. The correspondence between the populations may be caused by the number of foxes in the area, for example. The adage "correlation does not imply causality" is important in science.

## Sampling

Many phenomena studied in science have variation within the variables being studied. Humans, for example, vary in height and weight, behavior, income, susceptibility to disease, and many other factors. An experiment on one person, therefore, does not give results which are applicable to all people. On the other hand, to conduct an experiment on all people would be impractical. (Dear Granting Agency: I need enough money to collect and analyze blood from six billion people. . . .) A compromise used in science is to use a *sample*, a subgroup of a population chosen to be representative of the whole group. Statistical procedures (not discussed here, see a statistics text) are used to select an appropriate sample.

## Control Groups

The ball drop experiment is a simple one in which only two variables are involved: the density of the fluid and the time of drop. Most research involves more complicated situations in which many variables are involved and isolating the effect of just one is difficult. A way around this problem is to divide the study into test subjects and control subjects. Both groups are treated the same way, with the exception of the independent variable under investigation. This way, if the test subject results are different from the control subject results, then the variable under study is likely to be involved. To study the effect of a fertilizer on tomato production, plants can be put in identical soil, given identical amounts of water, and placed so that they receive the same amount of sunlight. Then some of the tomato plants are given the fertilizer and the rest are not given any. If the test group, those getting fertilizer, produces more tomatoes than the control group, those without the fertilizer, then it can be concluded that the fertilizer caused the improved yield. Without the control group, there is no way to guarantee that the fertilizer helped produce more tomatoes, since there is no group with which to compare results; there is no way to be certain that the fertilizer was the cause of the improved growth. It is easy to assume that the fertilizer did it, but such leaps of logic can lead to wrong conclusions. Proper use of control groups simplifies complicated problems and makes meaningful research possible in many circumstances.[21]

## Control in Experiments

Sticking with the ball drop experiment, every significant aspect of this test is under control by the scientist: nothing changes except the density of the fluid, and the scientist changes that. This is an example of a highly controlled experiment, which can often be achieved in laboratory studies. Quite often, however, the phenomenon under study does not lend itself to such strict controls. In most field experiments, for example, there are potentially significant variables that cannot be controlled. One may hypothesize that a predator fish population determines the population of a prey species in lakes and conduct an experiment to test it. This may involve measuring the populations of predator and prey over a period of time and then trying to find a relationship between the two. In this experiment, natural changes are monitored; there is no experimental control. A degree of control can be achieved by artificially manipulating the predator population (releasing new individuals if there are too few or trapping and removing some if there are too many), but it would be difficult to keep the population precisely at the desired level. In addition, there are other variables that can affect prey populations, such as disease and food supply, which make it difficult to determine the extent to which predator populations cause changes in prey populations. An alternative experiment would be to bring predators and prey into a large tank where they can be monitored more carefully, and food and disease are controlled, as well. This would make the study more like a laboratory experiment, with higher levels of controls on all variables. While there are definite advantages to such an approach, the scientists must keep in mind that the fish are no longer under natural conditions and the results found may not apply to real-world situations. In much of science there is a trade-off between studying actual conditions with poor experimental control and achieving high degrees of control under artificial conditions. Both have strengths and weaknesses in terms of the reliability of the results.

The phenomena studied in most of physics and chemistry are such that they can be done realistically under carefully controlled laboratory conditions, and such problems are avoided. This is not the case, though, with much of sociology and astronomy, for example, where experimental control is difficult under realistic situations. The phrase "hard sciences" is sometimes used to characterize such fields as chemistry and physics, which can maintain considerable control in their studies and have strong confidence in the results. The alternative is "soft sciences," where the subject matter is such that controlled experiments are much more difficult to do, reducing the reliability of the results. Especially in the soft sciences, a difficult but important part of experimental design is finding the best combination of control and realistic conditions for the problem at hand.

The reliability of an experiment increases if the test is repeated and consistent results are obtained. Even under highly controlled conditions, there is a possibility that the outcome of an experiment is due to an unknown error. Replication of the test helps check for spurious results. Experiments done in poorly controlled situations require more replication of tests because, in addition to unidentified errors, the uncontrolled variables cause variation in the results. Typically in such cases, the results of many experimental runs are averaged, and these mean values are used in the analysis.[22]

*Experimental Bias*

The results of an experiment can be distorted due to various forms of bias. *Sampling bias* occurs when the sample is unrepresentative of the population under study. For example, a poll was taken before the 1936 U.S. presidential election which incorrectly suggested that Franklin Roosevelt would lose. The problem was that the people polled were selected from telephone directories, though many of Roosevelt's supporters were too poor to own telephones. The sample of voters, therefore, did not reflect the population of voters.[23] Often, care must be taken in collecting data during an experiment because the measurement device itself can change the phenomenon under study. If the velocity of water flowing through a pipe is measured with an instrument placed in the pipe, the presence of the device will disrupt the flow of water and change the speed. Care must be taken to correct for such biased conditions. A deer may not act naturally if it knows that a person is nearby watching it, even if the person happens to be a biologist studying deer behavior. The same holds for people being studied by anthropologists.

*Psychological bias* in experiments has several forms. *Subject bias* involves distortion of the results by the unintended actions of the subjects of the study. In testing the effectiveness of a drug, for instance, typically one group is given the drug and another is given a *placebo*, a pill known to have no physiological effect. In a *blinded study*, the subjects do not know if they are taking the drug or the placebo. It is not uncommon, though, for people taking the placebo to improve medically; their improvement, however, is due to psychological factors, not from the contents of the pill they took; such effects must be considered in evaluating the results of the study.

*Experimenter bias* results when the researcher unknowingly distorts the results of the study. While scientists are expected to be disinterested in the results—not favoring one outcome over another—in practice, this is difficult to achieve. In a blinded drug study, as mentioned previously, the subjects do not know if they are taking the drug or the placebo. If the researcher does know, though, and he or she expects the drug to be more effective than the placebo, then when evaluating the patient's condition, he or she might subconsciously bias the results. If one patient with the drug and one with the placebo have equal improvement, the researcher subconsciously may rate the improvement in the drug patient higher because that is what he or she expects to find. This problem is eliminated in *double-blind studies*, where the researcher does not know what kind of pill each patient takes (records are kept, but not given to the researcher until after the data have been collected).

A famous case of experimenter bias is the "discovery" of N rays in 1903. René Blondlot was a well-respected French physicist at the University of Nancy. In experimenting with the newly discovered X rays, Blondlot found a new type of radiation, which he called N rays, named for his university. Without going into details of the experiment, N rays were detected by seeing a very faint light on a screen. After Blondlot published his findings, many other scientists reported on N rays as well. R. W. Wood was skeptical of this new form of radiation and visited Blondlot's laboratory. While Blondlot was repeating his experiments for Wood, Wood secretly removed a prism which was necessary to project the rays onto the screen. Blondlot got the same results that he would have gotten with the prism in place. It became clear that Blondlot was seeing N rays if he expected them to be there and not seeing them if he did not expect them, while in fact they never existed. The detection of N rays required faint light to be seen, and at the limits of detection, it is easy to be mistaken.[24] Scientists must be careful not to fool themselves in such ways.

*Subconscious signaling* leads to distorted results in some research situations. This is sometimes referred to as the "Clever Hans Phenomenon." Clever Hans was a horse in the early twentieth century who was believed to be able to do mathematical calculations: he would tap out the answers with one of his hooves. Skeptical scientists could find no evidence that it was a trick and declared his abilities genuine until it was discovered that

Hans's owner would unknowingly relax his muscles or nod his head in an almost imperceptible way when Hans had reached the correct number of hoof taps. The horse was not doing math, just watching his owner for very subtle clues which were followed by a sugar cube as a reward.[25] Wilson mentions an experiment similar to the one for N rays except that the experimenter was not supposed to know if a shutter was opened or closed before determining if a faint light could be seen. An assistant secretly opened or closed the shutter and then noted if it was open (writing *o*, for *offen*) or closed (writing *g*, for *geschlossen*). They discovered that the experimenter was subconsciously tipped off as to the shutter's condition by the length of time it took the assistant to make his note (*g* takes longer to write than *o*), and this affected the results. When the note-taking system was modified, the results changed.[26] Discovering sources of bias (or other errors in experimental methods), correcting them, and then repeating the experiments is common in scientific research; it can be aggravating and expensive, but at the same time, it can give the researcher pride in quality over expedience. Not correcting a known source of bias is an example of science badly done.

Reducing bias in research is generally done by eliminating as much as possible the subjective decisions made by the scientist. For example, rather than looking at a group of patients and then choosing which go into the drug group and which go into the placebo group, they should be put in the groups randomly. Having a machine determine if a faint light is present or not is better than having the scientist decide. While subjective decisions should be minimized in a study, they rarely can be eliminated totally. For example, when sampling the plant species in a research plot, a botanist might stretch a string over the ground and note the species present under the line. Where the string is placed should not be a conscious decision, because to look over the area and choose where to sample may lead to the researcher picking a location where certain species are found. He or she probably has a preconceived idea as to which species are likely to be found in the plot and may place the string, either consciously or subconsciously, so that those species are sure to be counted. This, of course, makes the sample unrepresentative. An objective way to sample is to choose the location of the string randomly. Even if the string is placed randomly, problems can arise. When sampling, for example, most plants will be obviously under or not under the line, but some will require judgment calls as to whether or not the edge of a leaf is under or not under the line. The botanist might deal with this problem by deciding beforehand that the first time such a situation arises, the plant is counted; the second time it is not, etc. Scientists must be aware of subjective decisions and be sure that they are not made in such a way as to bias the results.

## PERFORMING EXPERIMENTS

Experimental design creates a plan to follow when conducting an experiment to test a specific hypothesis. Ideally, this *experimental protocol* provides the detailed guidelines needed to do the work from start to finish. Typically, though, unexpected complications arise as experiments progress, and the protocol is changed. This situation is especially common in work involving new techniques, new equipment, or when the phenomenon in question is poorly understood. Even experienced scientists cannot anticipate all of the problems that can arise during research. A judgment call is needed when changing the experimental plan. Minor changes, those having little effect on the outcome, can be made readily and described only in the notebook. Significant changes, however, must be made carefully and should be reported in the resulting paper. If changes to the protocol alter the nature of the experiment itself, then the entire project should be started over with a revised experimental design, because the experiment is no longer addressing the objective of the study. When the study gets to the point of not having a plan to follow, and the experiment is made up as the scientist goes along, it is easy to be led astray and obtain misleading results. The details of experimental protocol vary widely, depending on the type of research done. All research, however, requires the reduction of measurement error and the taking of proper notes, which are the subject of the next two sections.

### Measurement Error

When doing experiments, continued vigilance is required to minimize erroneous measurements, which can arise in several ways.[27] *Systematic errors* are the same for all measurements. This can be due to incorrect instrument calibration, for example, where all readings are incorrect by the same amount. *Personal errors* are due to the judgments which are sometimes needed in making a measurement. If two scientists are using tape measures to determine the lengths of lion tails, they might not get the same measurements; one may tend to get higher values than the other. Even without the haste imposed by the threat of the lion eating them as they measure, there is judgment involved in reading the tape, and this can lead to error. Having several people measure the same objects is one way to check for personal errors. Making measurements on objects of known value is a way to check for systematic error.

*Unsystematic errors* can arise from a number of uncontrolled factors, each of which has a small effect on the results. Ideally, all factors that affect the variable under study are measured and included in the analysis, but this is not always possible or practical. Plant growth in a field can be studied, where the independent variable is the amount of water available to the plants. Slight variations in other variables, such as temperature, sunlight, nutrients, and the proximity of weeds, will cause scatter in the results. Another source of unsystematic error is the measuring equipment itself. For mechanical or other reasons, two measurements of the same object by the same piece of apparatus may give different values. If the cause of the variation cannot be eliminated, taking several readings and using the average is one way around this problem. Frequent calibration (testing of equipment to see if it is working properly) is important in research.

*Mistakes* are another source of error. These include writing down a wrong number (e.g., 26.2 instead of 2.62) or reading a scale incorrectly. Equipment malfunction can also provide mistaken readings. Simply staying alert when taking and analyzing measurements can minimize this problem.

Experimental design must include consideration of the reduction of measurement errors. When choosing ways to collect data, those methods least likely to produce error are preferred.

Another problem in research is sloppiness. Scientists are expected to design, conduct, and report their experiments carefully to improve the chance that the conclusions drawn will accurately reflect reality. Experiments done in a slipshod manner damage science because others will rely on a conclusion when such trust is not warranted.[28] The related problem of laziness is unacceptable, too. All of the work needed to complete an experiment must be done as specified and honestly reported.

## Note Keeping

Like reducing measurement error, proper note keeping is an important part of doing science. A well-kept notebook is a thorough record of a research project, from the thinking that went into planning the study to how the final results were obtained. The notebook is the scientist's record of her or his daily work, with all of the details that might be useful later. It contains the experimental setup, all of the data collected, an explanation of the analysis procedures used, and the results. It shows the problems encountered along the way and how they were overcome. Ideally, another scientist should be able to read a notebook and repeat a project completely. Another important part of a scientific notebook is that it is permanent. Erasing or removing information entered in a notebook is not allowed.

> A laboratory notebook is one of a scientist's most valuable tools. It contains the permanent written record of the researcher's mental and physical activities from experiment and observation, to the ultimate understanding of physical phenomena. The act of writing in the notebook causes the scientist to stop and think about what is being done in the laboratory. It is in this way an essential part of "doing good science."
>
> Howard Kanare, 1985[29]

Scientific notebooks serve several functions. First, the scientist needs a complete record of her or his work in order to keep track of the details when writing up the study. All experienced scientists have learned the hard way that all of the important details cannot be stored in the head, especially since all that is important may not be known until long after the data are collected, and the details may not be needed until years later. Second, another scientist may want to replicate a study, and following detailed notes is the best way to do that. Third, in the event of an important discovery or patent dispute, the notebook can show when a scientist was working on a particular study. (Some scientific research leads to lucrative technology, and laboratory notes can be an important part of the patenting process.) Fourth, a scientist's work may be questioned. If the results contradict the results from other studies, evaluating the procedures used may shed light on why the different results were obtained. More seriously, a scientist may be accused of misconduct (see Chapter 5), and notebooks can be used to show what was actually done in the study. Altering notebooks is a serious offense in science. The notebook must reflect what was actually done.

The part of science made public typically is the published research paper. This provides a justification for doing the study, an overview of what was done and how, and a presentation and interpretation of the results. These papers, though, are highly selective in what information they include; rarely can a study be reproduced precisely based solely on the published article.[30] The notebook, however, should have all of the potentially relevant information needed to reproduce the project.

That notebooks are a permanent record of everything done in a study is an ideal situation. Some scientists are superb notebook keepers, while most see room for improvement. It is a skill that scientists should be taught early and work on diligently throughout their career.

## ANALYSIS

The information collected in an experiment must be analyzed before conclusions can be drawn. The hypothesis gives a prediction of the outcome of the experiment, and the analysis determines if the prediction is accurate or not. Sometimes the results are so obvious that the hypothesis can be accepted or rejected easily, but it is common for scatter in the data (meaning variation from a clear pattern) to require a judgment call as to whether or not the hypothesis is supported by the experiment. Statistical procedures (not discussed here—see a statistics text for details) are used to find out if the hypothesis is supported or rejected by the experimental data.

Often the test is set up using a *null hypothesis*, which is the opposite of the real hypothesis. Better stated, the null hypothesis predicts the results which would be obtained if the real hypothesis is not true. To use a simple example, say that the hypothesis is that variable A affects variable B. In this case, the null hypothesis is that B is independent of A (i.e., A has no systematic effect on B). Analysis of the data then determines if the null hypothesis is or is not supported by the experimental results; if it is, then the hypothesis can be rejected. If the null hypothesis is not supported, then the true hypothesis can be accepted, at least from a statistical point of view. As discussed in Chapter 2, a hypothesis can never be proved true, but it can be shown to be wrong. If the results of a study indicate that the null hypothesis should be accepted, then the actual hypothesis can be rejected. If the null hypothesis is rejected, then more confidence can be placed in the true hypothesis.

Statistically speaking, there are two major errors that can be made in the interpretation of experimental results. A *Type I error* is to reject a hypothesis that is correct, and a *Type II error* is to accept a hypothesis that is wrong. An important part of experimental design is to plan an experiment that minimizes the likelihood of both kinds of error. This is one reason why single studies in science should not be used as conclusive evidence for some theory. Error is an integral part of the process.

An important part of statistical analysis is assigning a *significance level* to the test, meaning an acceptable level of risk that the conclusion might be wrong. A scientist may decide that 95 percent confidence is reasonable for a particular study, meaning that there is a 5 percent chance that the results were due to factors other than those accounted for in the experiment (e.g., random variation). The scientist can then determine statistically if the data support the hypothesis within that level of potential error. A statement such as "The results are significant at the 0.05 level" means that the statistical analysis suggests that there is less than a 5 percent chance of committing a Type II error. There are plenty of other ways to be in error, such as bad data or improper experimental design, so a *significance level* only refers to the statistical testing of the data. How much statistical error to accept, be it 5 percent or 1 percent or 10 percent, is a decision the scientist makes, depending on the nature of the problem. A study on the usefulness of a drug where being wrong can mean that people die requires a very low chance for error, ideally much less than 1 percent. Simply put, more scatter in the data means a lower significance level, so more data are needed to establish a clear pattern and get a desired level of significance. Obviously, getting more data adds to the time and cost of doing scientific studies.

In most experimental work, not all data are used in the final analysis. Proper experimental design and care in data collection can reduce the amount of bad data, but rarely can it be eliminated entirely. The problem confronting the scientist is how to decide which data to reject and which to use in the analysis. When collecting data, sometimes it is obvious that equipment malfunctioned or that a meter was read improperly; in such situations, the scientist makes a note to reject those data points. Often, though, problem data are not noticed until after they have been collected. When plotting data, for example, one point may be significantly different from the rest. The question then is why is it so different? Is it a mistake, or does it reflect the real situation, just in an unexpected way? The scientist can check his or her notes and see if a problem can be detected. If possible, that experimental run can be replicated to see if similar results are obtained. The scientist can also check the situation in more detail to see if there is a good explanation that had not been thought of before. It is wrong in science to reject data simply because they do not fit an expected pattern; there must be a good reason to remove a data point. That same good reason must be used to reject all data with the same potential problem, those that do not support the hypothesis and those that do. Which data to keep and which to reject is a personal decision by the scientist, though each decision to reject should be mentioned in the notebook, and the scientist should be prepared to defend that decision in front of a group of his or her peers. (Rarely is such a defense needed, but it is useful to imagine explaining experimental decisions to other experts. If they would not agree, then perhaps it is not the correct choice.) It may be tempting to use only the data which "look right," but it is bad science. Experiments are done to test hypotheses, not to confirm them; there is a fundamental difference between the two goals.[31]

---

There is, for example, the famous case of the discovery of argon by Lord Rayleigh, which arose because of a discrepancy between the density of nitrogen prepared from air and that of a sample produced chemically. It would have been easy for him to reject his result as having been caused by some mistake.

E. Bright Wilson, Jr., 1952[32]

---

## RESULTS

Analysis of the data collected in a scientific study provides the results of the study so that it can be determined if the hypothesis is supported, is not supported, or the results are inconclusive. Statistical procedures determine which of these answers is to be chosen.

It is important to understand that negative results, where the experiment does not support the hypothesis, are valuable. Knowing what is not correct can be as useful as knowing what is correct. Ideally, if the hypothesis is worth testing and the test was done well, then it is an important study and worth sharing with the scientific community. Whether the results are positive or negative does not determine the value of the research.

In reality, however, it often is easier to get a study with positive results published, so scientists may feel pressure to get results that support the hypothesis. This attitude is fine if the scientist gets negative results, then uses insight from that experiment to revise the hypothesis and do a new experiment. It is wrong, though, if positive results are obtained by altering the data (see Chapter 5 on scientific misconduct) or by altering the hypothesis after the experiment is completed so that it appears that positive results were achieved. Such a "*post hoc* hypothesis" is wrong because the experiment was not designed to test it, and the conclusions therefore might be erroneous. A new hypothesis requires starting over with a new experimental plan.[33] It is desirable to get "clean" results that clearly support a hypothesis, but to get clean results by altering the hypothesis is bad science. Good science is done when a careful plan is followed as closely as possible to test a specific hypothesis.

## DISCUSSION OF RESULTS

Hypotheses are subjective, and experiments are designed to be objective. The discussion of the results brings science back to the subjective. From a statistical point of view, the hypothesis is accepted or rejected, but what does this mean? Here the scientist attempts to put the results in perspective and gives her or his opinion on the significance of the study and how it fits into established knowledge. Also, the experience of doing the study provides insight that can benefit other scientists. Answers to a variety of questions can be presented. Does this study push the theory into new territory, or does it help confirm what we have already found? Should we consider changing how we think about the topic? Are there alternative explanations for the results? What other hypotheses could be tested? How could this hypothesis be tested in different ways? What was learned in doing the experiment about the strengths and weaknesses of the methods used? How reliable are these results? How could the study be done more effectively? Is the topic worth pursuing, or did this study show that it is a dead end? In answering these questions, care should be taken in making clear what is based on strong evidence and what is pure speculation. Speculation is important and useful in science but should never be confused with well-grounded information.

Two additional topics important to research have not yet been discussed. These are models and nonexperimental research.

### MODELS

Modeling is a useful tool in science. A scientific model is a simulation of a real thing (though a more precise definition is that models represent theories and the theories represent the real world). Modeling is used to overcome some of the limitations of studying the actual phenomena. Some things are too small or too large or too far away or too complicated to study only in their real situations, and models allow scientists to learn about them under artificial conditions. Some models are highly accurate representations of reality, while other models are quite inaccurate. The difference is largely due to the theoretical understanding of the phenomena under question.

Models come in a variety of forms. *Hardware models* are physical objects used to represent the real thing. In studying the aerodynamics of cars, for example, a small version of the car (a scale model) can be placed in a wind tunnel. In this case, the car is modeled and so is the wind. In the wind tunnel, the car is stationary, and the wind is carefully controlled. To study wind flow around a car under real conditions is difficult because the car has to be moving and the natural wind changes much more than in a wind tunnel. A crucial part of such a study is determining how the model differs from the real situation and, therefore, how applicable the results are to the real world. The structure of DNA molecules was determined by the use of a hardware model. James Watson and Francis Crick took the information they knew about DNA and about how the relevant atoms could be arranged and then proceeded to put together a molecular puzzle using these rules. The resulting pattern was a double helix. They then used other means to confirm that the double helix was the actual shape.[34]

Animals are often used as models in biological and medical research. Much has been learned about genetics by studying fruit flies, for example. This is not because fruit flies are inherently more interesting to geneticists; they simply reproduce quickly and are relatively easy to study, so much information about genetic change over many generations can be gained in a short period of time. With a clear understanding of some genetic phenomena in fruit flies, it is easier to find out if the same can be found in other types of animals. Another example is the use of mice in medical research. For ethical reasons, much medical research cannot be performed on people (more on this in Chapter 5). Therefore, animals are often used as models of humans. Studying cancer in laboratory mice, for example, has taught us much about cancer in people. While mice are similar to people in many ways, there are differences, and researchers must then determine if the results found for mice are applicable to humans.

---

People have a basic misunderstanding of the mouse. They get upset that the exact pathophysiology might be different, but a mouse is not a total mimic. A mouse is a discovery tool, a device, to understand the molecular pathways that underlie disease processes. You can't do a lot of types of experiments on humans. You can't order people to mate.... We have to turn to models.

Kenneth Paigen, 1998[35]

---

*Software models* simulate using mathematical equations to represent the phenomena under study. For example, the motion of Earth and the Moon can be represented mathematically using Newton's laws of motion, and this model can be used to predict when eclipses will occur. Global climate models simulate Earth's atmosphere using equations to represent the interactions between the many components of the air (temperature, clouds, wind, etc.). Because the atmosphere is so complicated, the models are by necessity highly simplified versions of reality. Even so, they require some of the most powerful computers in the world to run them.

Modeling is a fundamental part of science. Much of what we learn comes, at least in part, from models. However, a model is only useful if its limitations are clearly understood. A model is a model—it is not the real thing. Whenever possible, the results from a model need to be checked with the actual phenomena under study. A successful cancer treatment for rats may not be as helpful to humans. Wind flow around cars on a highway may be different from flow around scale models in a wind tunnel. Global climate models may be too simplified to accurately reflect the actual climate. A model is better thought of as a hypothesis generator, not an answer giver. It tells us what to look for in reality and makes the search easier by suggesting likely routes. Science requires evidence from the real world in order to draw strong conclusions.

Sometimes, though, scientists are unable to check model results directly with the real thing. Our ability to do field work on the origin of the universe and on future events, for example, is limited at best. In such cases, model results may be the best results scientists can offer. The challenge then is to assess the reliability of the model. For example, Stephen Schneider discusses the limitations of global climate models in terms of their ability to predict future climates. (Such predictions are important in assessing the human impact on climate, its future consequences—say over the next century—and what we can do to mitigate the problem.) While we cannot truly know if the predictions are correct until the time the prediction comes true, we can get a better idea of the models' accuracy by using them to predict known situations. Can, for example, the models correctly predict the present climate? Can they predict the climate at times in the past when we know the situation was different, such as during ice ages? If the models do well in predicting known situations, then we can have more confidence in their ability to predict unknown ones. This does not guarantee that the predictions of future climates are accurate, but it makes it more likely that they are.[36] In making scientific predictions, uncertainty is reduced as much as possible, and that uncertainty is always kept in mind when using the results.

Much in science is so complex that truly accurate models are not to be expected. Many researchers would be happy if their scientific models simulated reality better than fashion models reflect the way typical people look when wearing clothes.

### NONEXPERIMENTAL RESEARCH

While most scientific research involves hypothesis testing, not all does, and not all research is experimental. Some studies are done to answer questions rather than test predictions. For example: What can we learn about past climates by analyzing ice cores from Greenland? What was the population of Mayans at the height of their civilization? What tree species exist in the Amazon rainforest? What patterns of nucleotides make up human genes? What is the surface of Venus like? In many areas of science, topics are poorly understood, and such basic information must be collected and analyzed in order to provide the background information needed to develop theories. Careful observation and description are needed before any science can develop. For example, the idea of continental drift (later developed into the theory of plate tectonics) began when reasonably accurate maps of the continents became available, and it was seen that the continents could be pieced together like a giant jigsaw puzzle. Without the Age of Exploration to provide the information needed to make the maps, no one would have thought of the theory. All areas of science go through an age of exploration, which merges with an age of theory development, and the two overlap and interact considerably. Some disciplines, like optics, are mostly past the exploratory stage, but others, like paleontology, still need much exploratory research.

Nonexperimental research is crucial in science. While controlled experiments often are the most effective way to advance scientific knowledge, not all topics lend themselves to that type of research. It is the topic at hand that determines if experiments are the best way to do the re-

scarch. Of course, any type of research requires the same rigor in the collection and analysis of information and in the presentation of the work.

Scientific research, both experimental and nonexperimental, has greatly
improved our understanding of the universe, especially much of what
happens on this small part of it we call Earth. Research, however, must
be done creatively and carefully to better our theories and the evidence
we use to evaluate them. It is crucial that studies be done properly, lest we
fool ourselves with misplaced confidence.

Research, by itself, is not enough to advance knowledge. The research
has to be evaluated by a larger group of scientists, who seek to weed out
the good from the bad in science. This is the major undertaking of the
community of scientists, which is the subject of the next chapter.

### PROBLEMS

1. "There is no need for hypothesis generation to be a logical
   process." Discuss this statement.

2. Why is serendipity in science not solely a matter of chance?

3. Think of a scientific study, either real or hypothetical. What
   dependent and independent variables are involved?

4. "Correlation does not imply causality." Discuss this statement.

5. Why are samples used in research? What is meant by "representative sample"?

6. You have a cold and take a new cold medicine. Two days later,
   your cold is gone. What can you conclude about the effectiveness
   of that medicine? (*Hint:* Think of control groups and samples.)

7. What is the trade-off between laboratory and field studies in such
   "soft science" subjects as sociology and geology?

8. How do double-blind methods reduce the chance of bias in a
   study on the effectiveness of a new drug?

9. Why are notebooks valuable in science?

10. "Erroneous conclusions are always a possibility in scientific research." Discuss this statement.

11. Why are *post hoc* hypotheses dangerous in science?

12. What are the strengths and limitations of models in science?

## NOTES

1. Saxe, John Godfrey, 1869, *The Poems of John Godfrey Saxe*, Boston: Fields, Osgood, & Co., p. 259–61. (Saxe's poem is a version of an Indian folktale.)

2. Using maps as a metaphor for science is not new. See, for example, Turnbull, David, 1993, *Maps Are Territories: Science Is an Atlas*, Chicago: University of Chicago Press.

3. Wilson, E. Bright, Jr., 1952, *An Introduction to Scientific Research*, New York: McGraw-Hill Book Company, p. 1–6.

4. This example was first used in Beveridge, W. I. B., 1950, *The Art of Scientific Investigation*, New York: W. W. Norton & Company, p. 41.

5. Ibid., p. 56.

6. Holton, Gerald, 1996, "On the Art of Scientific Imagination," *Daedalus* 125, no. 2: 183–208; quote on p. 189.

7. Simplified from Bunge, Mario, 1967a, *Scientific Research I: The Search for System*, New York: Springer-Verlag, p. 243–47.

8. See Root-Bernstein, Robert Scott, 1989, *Discovering*, Cambridge: Harvard University Press.

9. Beveridge, W. I. B., 1980, *Seeds of Discovery: A Sequel to The Art of Scientific Investigation*, New York: W. W. Norton & Company, p. 19, 120–21. See also Asimov, Isaac, 1982, *Azimov's Biographical Encyclopedia of Science and Technology: The Lives and Achievements of 1510 Great Scientists from Ancient Times to the Present Chronologically Arranged*, 2d rev. ed., Garden City, New York: Doubleday & Company, p. 30.

10. Remer, Theodore G., ed., 1965, *Serendipity and the Three Princes: From the Peregrinaggio of 1557*, Norman: University of Oklahoma Press, p. 7.

11. Beveridge, 1980, p. 20–33.

12. Medawar, Peter, 1996, "Lucky Jim," in *The Strange Case of the Spotted Mice and Other Classic Essays on Science*, Oxford: Oxford University Press, p. 99. (This essay originally appeared as a book review of J. D. Watson, *The Double Helix*, *New York Review of Books* 28 (March 1968).

13. Beveridge, 1950, p. 33–34.

14. Bunge, 1967a, p. 272–78. See also Strahler, Arthur N., 1992, *Understanding Science: An Introduction to Concepts and Issues*, Buffalo: Prometheus Books, p. 55–57.

15. Bunge, 1967a, p. 247.

16. Beveridge, 1950, p. 81–91.

17. Barber, Theodore X., 1976, *Pitfalls in Human Research: Ten Pivotal Points*, New York: Pergamon Press, Inc., p. 16–18.

18. Bunge, 1967b, *Scientific Research II: The Search for Truth,* New York: Springer-Verlag, p. 162–69.

19. Shapere, Dudley, 1982, "The Concept of Observation in Science and Philosophy," *Philosophy of Science* 49: 485–525. See also Bunge, 1967b, p. 153–69.

20. Brownowski, Jacob, 1978, *The Origins of Knowledge and Imagination,* New Haven: Yale University Press, p. 58–59.

21. Wilson, 1952, p. 40–43.

22. Ibid., 46–48.

23. Ibid., p. 39.

24. Langmuir, Irving, 1989, "Pathological Science," *Physics Today* 42: 36–48. (Transcribed and edited by Robert N. Hall from a 1953 recorded lecture by Langmuir.)

25. Sagan, Carl, 1979, *Broca's Brain: Reflections on the Romance of Science,* New York: Random House, p. 50–51. See also Randi, James, 1995, *The Supernatural A–Z: The Truth and the Lies,* London: Headline Book Publishing, p. 65–66.

26. Wilson, 1952, p. 45.

27. Ibid., p. 232–34.

28. See Lederberg, Joshua, 1995, "Sloppy Research Extracts Greater Toll Than Misconduct," *The Scientist* 9, no. 4 (February 20): 13.

29. Kanare, Howard M., 1985, *Writing the Laboratory Notebook,* Washington, D.C.: American Chemical Society, p. 1.

30. For a discussion of how the scientific paper does not represent the process of science, see Medawar, P. B., 1963, "Is the Scientific Paper a Fraud?" *The Listener* 70, no. 1798 (September): 377–78.

31. Wilson, 1952, p. 256–58.

32. Ibid., p. 256.

33. Bailar, 1986, "Science, Statistics, and Deception," *Annals of Internal Medicine* 104: 259–60.

34. See Watson, James, D., 1968, *The Double Helix,* New York: Signet Books.

35. Quoted in Lewis, Ricki, 1998, "How Well Do Mice Model Humans?" *The Scientist* 12, no. 21 (October 26): 10–11.

36. Schneider, Stephen H., 1987, "Climate Modeling," *Scientific American* 256: 72–80.

CHAPTER 4

# THE COMMUNITY
# OF SCIENTISTS

Thus at the vital point in his life work [doing research] the scientist is cut off from communication with his fellow-men. Instead, he has the society of two, six, or twenty men and women who are working in his specialty, with whom he corresponds, whose letters he receives like a lover, with whom when he meets them he wallows in an orgy of talk, in the keen pleasure of conclusions and findings compared, matched, checked against one another—the pure joy of being really understood.

The praise and understanding of those two or six become for him the equivalent of public recognition. Around these few close colleagues is the larger group of workers in the same general field. They do not share with one in the steps of one's research, but they can read the results, tell in a general way if they have been soundly reached, and profit by them. . . . [If the scientist's close colleagues are impressed with his research, word will spread that his work is well done and trustworthy.] All told, including men in allied fields who use his findings, some fifty scientists praise him; before them he has achieved international reputation. He will receive honors. It is even remotely possible that he might get a raise in salary.

*Oliver La Farge, 1942*[1]

Scientists cannot exist in isolation and effectively contribute to the advancement of knowledge. Science is a social institution. It seeks to balance the sometimes competing themes of cooperation among scientists in the sharing of knowledge and information and the competition for limited rewards. A healthy skepticism of the work of others is an integral part of the institution as well. As John Ziman states, ". . . scientific knowledge is the product of a collective human enterprise to which scientists make individual contributions which are purified and extended by mutual criticism and intellectual cooperation. According to this theory *the goal of science is a consensus of rational opinion over the widest possible field."*[2] As Robert Merton phrases it, "The institutional goal of science is the extension of certified knowledge."[3]

Chapters 2 and 3 outlined how scientific knowledge initially is gained. New findings and new ideas must also be presented to other scientists for evaluation. Knowledge accepted as valid by most of the scientists in a discipline is generally considered the most reliable. This chapter is concerned with the norms and activities which have developed to facilitate the search for consensus among scientists.

## SCIENTIFIC NORMS

In 1942, Robert Merton outlined the *scientific norms*, or the rules and expected behavior for the institution of science: communism, universalism, disinterestedness, and skepticism.[4] Later, Ziman added originality and organized the norms under the acronym CUDOS.[5] Together, the scientific norms (also called the *ethos of science*) provide a foundation of how the community of scientists is expected to operate. Exceptions to each of these ideals, of course, do occur.

### Communism

Scientific knowledge is the property of everyone. Newton discovered the laws of motion, but he did not own them. They were and continue to be available for all to use. Scientists have an obligation to publish their results, thus bestowing them to the community. Merton commented that "Newton's remark—'If I have seen farther it is by standing on the shoulders of giants'—expresses at once a sense of indebtedness to the common heritage and a recognition of the essentially cooperative and selectively cumulative quality of scientific achievement."[6] The "payment" given to scientists for their contributions is in the form of recognition and esteem. (Scientific communism should not be confused with the economic system of communism; in practice, the two have little in common. Some, like Ziman, prefer to use the term *communalism*.[7])

Scientific communism is not always practiced, however. Many scientists work in secrecy. Military research, for example, often is not disseminated to the scientific community to keep the resulting knowledge out of the hands of enemies. Likewise, research on computer system security typically is not made public. Most scientists working in commercial research do not publish their work in order to keep their findings from competitors. While exceptions to the communism norm exist, most basic and some applied research is offered freely to the scientific community.

## Universalism

The principle of universalism, as summarized by Ziman, is that "[t]here are no privileged sources of scientific knowledge."[8] A scientist's nationality, gender, race, religion, political persuasion (or hat size, for that matter) should have no effect on the way her or his work is judged. The only criteria by which the professional work of individual scientists should be judged is by their competence as scientists. Likewise, scientific statements should be judged on their own merits, and the personal characteristics of the scientists proposing them are irrelevant to the judging. Even "great" scientists need to have their ideas judged by evidence and logic, not their record of past accomplishments. Even Linus Pauling, one of the greatest chemists of the twentieth century, was not able to convince most medical scientists of his conviction that large doses of vitamin C are medically beneficial. He had insufficient evidence to support his claim.[9]

One of the great commandments of science is, "Mistrust arguments from authority." (Scientists, being primates, and thus given to dominance hierarchies, of course do not always follow this commandment.) Too many such arguments have proved too painfully wrong. Authorities must prove their contentions like everyone else.

Carl Sagan, 1995[10]

## Disinterestedness

Disinterestedness refers to a lack of bias or a lack of self-interest. (It is not the same as being *un*interested.) Scientists are expected to pursue their work mainly for the advancement of knowledge, not for personal gain or to advance a political cause. Most scientists do feel strongly about their work, its reception among the scientific community, and its potential for use in solving problems. However, the community of scientists has an obligation to see that the pursuit of knowledge is the primary goal, not self-promotion. Advancing one's career through unethical means is a serious offense in science, as discussed in Chapter 5.

## Originality

Scientific work should advance knowledge. Scientists are rewarded most for exploring new areas of research, providing new insights on old problems, and developing new techniques for research. Filling gaps in well-established knowledge and reproducing experiments to test their results are significant research endeavors as well. Scientific work that provides nothing new to our understanding is not regarded highly.

## Skepticism

It is the duty of the community of scientists to judge scientific work carefully and critically and to reject work that lacks merit. Scientists are expected to be critical of their own work and the work of others, as discussed in Chapter 6. In addition, skepticism is institutionalized in the form of peer review, which is described later in this chapter.

### INVISIBLE COLLEGES

Most scientists work at universities or research laboratories, but they interact extensively with scientists located elsewhere and, so, are also members of *"invisible colleges."*[11] Such communication networks transcend national borders and have formal and informal components. Formally, journal articles are the primary communication medium. Conference proceedings are another means of formally presenting research results. Somewhat less formal are talks given at scientific meetings or as invited lectures. Such talks provide the scientist an opportunity to inform others of the nature of recent research and to receive feedback from peers before submitting results for publication.

Less formal communication is very important in science as well. Groups of scientists who share a research topic often are loosely organized into networks. Before submitting a paper for publication, it is common to send it to colleagues for comments on how to improve the paper. After papers are published, copies (called reprints) typically are sent to others who share similar interests. Even before official publication, it is common for copies of papers (called preprints) to be sent to colleagues so that the results can be used sooner by others. Scientific conferences have formal components, such as paper presentations, but a gathering of scientists with similar interests also has an important informal side, with much sharing of information (research findings, techniques, funding opportunities, job seeking, gossip, etc.) going on in hallways, during meals, and at "social hours." E-mail discussion lists are also used to disseminate information and discuss topics of interest to a particular group.

## PEER REVIEW

Members of invisible colleges are involved in the evaluation of each other's work. Most scientific research is judged formally through the process of *peer review*. When a research paper is submitted to a journal for publication, the editor sends it to several *peer reviewers*. These are scientists who are chosen for their knowledge of the subject matter, techniques used, and any other relevant information needed to judge the quality of the research. The job of the reviewer is to make sure that the research was done using appropriate techniques, that the conclusions are justified, and that it is a useful contribution to science worthy of being published. Reviewers commonly rate papers as acceptable, unacceptable, or conditionally acceptable. In the case of the latter, a list of changes is provided which will make the work acceptable in the reviewer's opinion. The editor uses the advice of all of the reviewers to decide if the paper is to be published, not published, or sent back to the scientist for further work. This process is designed to ensure that only the best papers are published, especially in the most prestigious journals, where the standards of acceptability are expected to be quite high.

A similar procedure is used to evaluate research grant applications. Typically, a scientist submits to a granting agency a proposal in which the research is outlined in terms of why it is important, how it will be done, how much money it will cost, and why the researcher is qualified to undertake the project. An "in-house" panel of experts evaluates all of the proposals submitted and, with the advice of outside reviewers, decides which of the proposals are to be funded and how much money each researcher will receive. In the case of proposal reviewing, the intention of the peer review system is to achieve the greatest amount of scientific advancement from a given amount of money.

The identity of peer reviewers typically is kept secret and often the author of the article is unknown to the reviewer as well. Scientific papers are to be judged on the merits of the research presented and not on the reputation of the scientist, which is an expression of the *universalism* norm discussed previously. Personality differences, professional jealousy, and competition for limited research funds are not supposed to be part of the peer review process.[12]

In practice, however, the peer review process is not perfect. Often the reviewers are competitors of the authors (or grant proposers) for research money and prestige, and even when anonymity is used, factors other than scientific merit sometimes come into play. Peer review also can be stifling to ideas that contradict prevailing theories. For example, in 1963, Lawrence Morley submitted a paper in which he outlined a crucial idea that led to the acceptance of plate tectonics theory, the most important advancement in the earth sciences in the twentieth century. Morley's paper was rejected by two leading journals. One reviewer commented that ". . . such speculation makes interesting talk at cocktail parties, but it is not the sort of thing that ought to be published under serious scientific aegis."[13] (As an aside, Morley's hypothesis was independently developed by two other geophysicists at the same time, Fred Vine and Drummond Matthews. Vine and Matthews submitted a paper for publication outlining the hypothesis after Morley submitted his, but their paper was published, and they are usually given credit for the theory, which rightly should be shared with Morley.)

### REWARD SYSTEM

Scientists are rewarded for their contributions in many ways. Sometimes the rewards are monetary, as in the case of higher salaries and profits from patents derived from research, but most rewards are in the form of *peer recognition*. Perhaps the most impressive form of scientific recognition is *eponymy*, or having something named after the scientist. Robert Merton has ranked eponyms, and at the top of the list is having a scientific age named for the scientist, as in the Newtonian Epoch in physics or the Freudian Age in psychology. Next in line is to "father" a discipline; for example, Robert Boyle is the "Father of Chemistry" and Georges Cuvier is the "Father of Paleontology." (It is strange that Marie Curie is not referred to as the "Mother of Radioactivity.") Eponymy continues with having laws, theories, etc. named for the scientist, like Boyle's law in chemistry and "Keynsian Economics." Advancements in a discipline are sometimes rewarded with the use of the scientist's name, such as Parkinson's disease and Eustachian tubes in medicine. Other eponymous uses are naming units of measurement for scientists (e.g., kelvin, volt) and using scientists' names in species names and geographical features. For example, the various species of finch found on the Galapagos Islands are collectively known as "Darwin's Finches," one of which has the Latin name of *Geospiza darwini*, which, of course, is found on Darwin Island.[14] (Merton suggests that eponymy can go too far and gives the example of Father Johann Dzierson, the "Father of Modern Rational Beekeeping.") While eponyms are commonly encountered in science, a very small percentage of scientists are rewarded in such a way.[15]

Also relatively rare in science is the awarding of prestigious prizes and admission into honorary academies. The most famous of the scientific prizes is the *Nobel Prize*, which is given to the scientists who make the most important contributions in certain fields of science (chemistry, economics, physics, and physiology and medicine). Most disciplines have prizes which are awarded to individuals for important contributions. Many countries have honorary scientific societies that admit only "important researchers," such as the Royal Society (in the United Kingdom), the Indian National Academy of Sciences, and the National Academy of Sciences (United States).

Scientists are rewarded through their employment, too. It is a sign of prestige to work at such elite universities as Oxford, Harvard, or most obviously, _____,* or an elite research laboratory, such as Scripps Institution of Oceanography. "Better" scientists are supposed to be rewarded for their contributions by being promoted in their jobs

---

*To minimize anger directed at the author, fill in the name of your favorite university. Print legibly.

more rapidly than their less productive peers. Unfortunately, "productivity" often is measured by the numbers of papers published and the amount of research grant money obtained and not by the significance of the contributions to the advancement of knowledge. Counting papers and grant money is easier than evaluating such contributions.

Publishing research findings and obtaining research grants are also forms of scientific reward. The peer review process is designed to reward scientists for good work by "weeding out" the less worthy research. The "best" scientists are expected to be those with papers in the most prestigious journals and grants from the most prestigious funding agencies.

Peer recognition comes in the form of citations in research articles, too. In papers, the work being reported is usually put in context with what already is known about the topic, with relevant and important papers cited in the text and listed in the reference section. Presumably, important research is cited more frequently than lesser studies (though sometimes papers are cited as examples of bad science).

## BECOMING A SCIENTIST

Most small children are budding scientists with a wonderful curiosity about the world around them and an insatiable quest for understanding it. Unfortunately, such enthusiasm rarely lasts through childhood. For some, though, the quest for knowledge is maintained, or rekindled, and they choose careers in science. Training in science varies both between and within countries, but the following serves as a brief overview. As undergraduates in college, students learn mostly general (or "textbook") knowledge of one or more scientific disciplines, along with various tools (like mathematics and laboratory methods), communication skills (both written and verbal), and other topics associated with a liberal arts education, such as the arts, literature, and history. In graduate school, students become more specialized and study "frontier" more than textbook research, they begin to conduct original research, and they learn specialized research techniques. Also in graduate school, students often enter invisible colleges by becoming involved in scientific organizations, attending conferences, and presenting and publishing their research. The *doctoral dissertation* is a piece of research done by the student which is expected to be a significant and original contribution to knowledge. It gives the student experience in conducting a major research project while still under the guidance of faculty members and also is used to judge the student with respect to his or her potential as a scientist.[16]

After earning a *Ph.D.* (Doctor of Philosophy) in their discipline, most scientists seek employment in universities (where teaching is expected in addition to research), research laboratories, or commercial firms. Scientists often work for one or more years as postdoctoral fellows (usually called *postdocs*) before entering permanent employment. Postdocs conduct original research, though usually in association with a more senior scientist. Scientists who are employed as university faculty are given a probationary period (five years is common in the United States) during which they must show their effectiveness as scientists and teachers. At the end of this probationary period, if their work is "good enough" for that university, then they are granted *tenure*, which is intended to guarantee *academic freedom*, or protection from being fired for political reasons. Academic freedom is important because knowledge grows best in societies that allow researchers in the sciences and other scholarly endeavors to espouse unpopular views in teaching and research without the fear of job loss.

Through the process of becoming scientists and throughout their professional careers, researchers gain experience that helps them work as part of the community of scientists. This is true in the formal aspects of the community, as in obtaining funding, publishing research findings, and receiving awards from peers, and the informal parts, such as where jobs will become available for their students and how to find appropriate research field sites without too many mosquitoes.

### PROBLEMS

1. John Ziman said, ". . . the goal of science is a consensus of rational opinion over the widest possible field." Discuss this statement.

2. How do each of the scientific norms benefit the search for scientific knowledge?

3. Find, for a particular scientific discipline, a national or international organization, a journal, and an e-mail list. List several

leading universities for this discipline. (*Hint:* None will be found in this book; asking someone in that discipline is a good way to find out such information.)

4.  What are the advantages and disadvantages of peer review in science?

### NOTES

1. La Farge, Oliver, 1942, "Scientists Are Lonely Men," *Harper's Magazine* (November): p. 652–59; quote is on p. 657.

2. Ziman, John, 1978, *Reliable Knowledge: An Exploration of the Grounds for Belief in Science*, Cambridge: Cambridge University Press, p. 2–3 [italics in original].

3. Merton, Robert K., 1973a, "The Normative Structure of Science," in *The Sociology of Science: Theoretical and Empirical Investigations*, Chicago: University of Chicago Press, p. 267–78; quote is on p. 270. (Originally published as Merton, Robert K., 1942, "Science and Technology in a Democratic Order," *Journal of Legal and Political Sociology* 1: 115–26.)

4. Merton, 1973a.

5. Ziman, John, 1984, *An Introduction to Science Studies: The Philosophical and Social Aspects of Science and Technology*, Cambridge: Cambridge University Press, p. 84–86.

6. Merton, 1973a, p. 274–75.

7. Ziman, 1984, p. 84.

8. Ibid.

9. Hager, Thomas, 1995, *Force of Nature: The Life of Linus Pauling*, New York: Simon and Schuster, p. 573–627.

10. Sagan, Carl, 1995, *The Demon-Haunted World: Science as a Candle in the Dark*, New York: Random House, p. 28.

11. Crane, Diana, 1972, *Invisible Colleges: Diffusion of Knowledge in Scientific Communities*, Chicago: University of Chicago Press. See also Price, Derek De Solla, 1986, *Little Science, Big Science . . . and Beyond*, New York: Columbia University Press, p. 56–81.

12. For more detail on peer review, see Grinnell, Fredrick, 1992, *The Scientific Attitude*, 2d ed., New York: Guilford Press, p. 76–82.

13. Glen, William, 1982, *The Road to Jaramillo: Critical Years of the Revolution in Earth Science*, Stanford: Stanford University Press, p. 269–311.

14. Grant, Peter R., 1986, *Ecology and Evolution of Darwin's Finches*, Princeton: Princeton University Press, p. 55.

15. Merton, Robert K., 1973b, "Priorities in Scientific Discovery," in *The Sociology of Science: Theoretical and Empirical Investigations*, Chicago: University of Chicago Press, p. 297–301. (Originally published as Merton, Robert K., 1957, "Priorities in Scientific Discovery," *American Sociological Review* 22, no. 6: 635–59.)

16. For more detail on graduate education, see Grinnell, 1992, p. 52–68.

# CHAPTER 5

# MISCONDUCT IN SCIENCE

Truth swings along a wide arc in the late twentieth century. In politics and popular entertainment, people accept deception as commonplace. . . . The trust that society places in science, however, traditionally assumes different standards, including assurances of authenticity and accuracy in all that science does or recommends.

. . . Forgery, fakery, and plagiarism contradict every natural expectation for how scientists act; they challenge every positive image of science that society holds.

*Marcel C. LaFollette, 1992*[1]

The most important personal characteristic of a scientist is not intelligence. Neither is it creativity or determination. The most important characteristic of a scientist is integrity. If a scientist is known to lack integrity, others will not believe his or her research findings, and that scientist's career may end. If the lack of integrity is not known by others, his or her work will be trusted when it should not be, and science may be led astray. As the quote from LaFollette shows, public trust in science can be undermined by dishonest scientists, which is significant because much of the financial support for science comes from the public through government research grants. In addition, as long as science is seen as an honorable profession, students with integrity will be attracted to it.

Science progresses because scientists present honest research in an honest way. There is a code of conduct in science dealing with the acceptable ways to do research, present research findings, and coexist with other scientists, which has been referred to simply as ". . . good manners in research."[2]

Scientific Commandments:

Be honest!
Never manipulate data!
Be precise!
Be fair with regard to priority and ideas!
Be without bias with regard to the data and ideas of your rival!
Do not make compromises in trying to solve a problem!

Hans Mohr, 1979[3]

If scientists behave in professionally unethical ways, science suffers. For example, inappropriately manipulated data reported in a scientific paper may mislead others in the discipline about the nature of their work and cause them to waste effort by testing poor hypotheses. Scientists need to know that the work of others is trustworthy, otherwise each scientist would have to do all experiments in order to have confidence in the results. In addition, when the public hears about unethical scientists, the image of all scientists may be tarnished. Unfortunately, the code of conduct cannot be written down in such a way that what is proper behavior is always obvious. Scientists learn how to make the decisions necessary for the conduct of good and ethical science. The purpose of this chapter is to introduce the topic of scientific integrity and present some of the ethical topics scientists must frequently consider during the course of their work.

The community of scientists is obligated to punish misconduct. Punishment ranges from being treated as unworthy of research grants to being driven out of science altogether. Unfortunately, scientific misconduct is not dealt with in a uniform manner by the institutions responsible for overseeing the actions of scientists. There is no "science court" in which all scientists accused of misconduct are tried. In some cases, universities and granting agencies prefer to ignore allegations of misconduct because they are embarrassing and can damage the institution's image.[4] Because the ultimate goal of science is the advancement of knowledge, which requires properly done research, such "sweep it under the rug" attitudes should be fought wherever they exist.

A panel of the (U.S.) National Academy of Sciences studied the issue of ethical problems in science.[5] They concluded that the most serious forms of misconduct involve fabrication, falsification, and plagiarism. The panel also listed other questionable research practices. Each will be discussed.

## FRAUD

*Fraud* in science generally involves fabrication or falsification. *Fabrication* is making up information used to justify a result or making up the result itself. *Falsification* is similar, but the data or results are changed inappropriately, such as improperly deleting data that do not support one's expected results.

Stephen Bruening was guilty of fabrication. Bruening studied drug therapy for severely mentally retarded children, and his published papers influenced how such children were treated. His research suggested a treatment of stimulants, though other scientists were recommending tranquilizers. Unfortunately, Bruening made up most of the data used to support his conclusions, thus potentially endangering helpless children in an attempt to further his career without doing all of the required work. His fabrication was discovered by Robert Sprague, who became suspicious of some of Bruening's data and investigated further. Sprague reported the misconduct to Bruening's funding agency and university. Eventually, Bruening pleaded guilty to submitting faked reports on federally sponsored research and is no longer a practicing scientist.[6]

V. J. Gupta authored or coauthored hundreds of papers, many using fossils to interpret the geological history of the Himalayas. It has been learned, though, that many of the fossils Gupta used as evidence were not collected in the Himalayas; many were not even from Asia. For example, fossils used to understand the geological past of Kashmir were really from New York. Most geologists working on related topics in the region unknowingly have relied on Gupta's conclusions in making their own interpretations. A tremendous effort is needed to determine which of Gupta's papers are trustworthy and which are not and purge the fallacious ideas which have permeated the discipline. If Gupta had done honest research, such damage control would not be necessary, and scientists could spend their time contributing to knowledge rather than cleaning up someone else's intellectual mess.[7]

While Bruening and Gupta were guilty of fabrication, physicist Robert A. Millikan practiced falsification. In 1913, Millikan reported on measurements that supported his theory on the value of the smallest natural electrical charge. While he stated that all of his experimental data were reported, study of his laboratory notebooks revealed that he presented only 58 out of a total of 140 experimental observations in the paper. Not surprisingly, the data he chose to publish were the ones that best supported his theory. Millikan received the Nobel Prize in 1923, partly based on these experiments. He died in 1953, and it was several decades later that his falsification was discovered and reported. In most experiments, there are runs which are not used in the final analysis because of problems with equipment or other factors. It is unacceptable in science, however, to delete runs just because they do not support a theory.[8]

## PLAGIARISM

*Plagiarism* is the use of someone else's words or ideas without giving proper credit. Such intellectual theft is not tolerated in science. There are many ways to plagiarize. Some cases of plagiarism are obvious, but many are judgment calls that scientists must learn to make. Using someone else's wording can be done only if it is presented as a quote and the original work is cited. The original work must also be cited if it is paraphrased. If some part of another work (data, results, methods, etc.) is used, even if the wording is significantly different, the source must be cited. Citations are not necessary, however, if ideas are widely known. For example, if "the earth revolves around the sun" is used in a paper, it is not necessary to cite the work of Copernicus. Everyone who reads the paper can be expected to know that the earth revolves around the sun and from where that idea came. However, if new findings about the details of how the earth revolves around the sun are used in a paper, then the source of that new information should be cited. When in doubt whether or not to cite, ". . . erring on the side of excess generosity in attribution is best."[9]

Perhaps the most notorious scientific plagiarist is Elias Alsabti, who "authored" dozens of papers on cancer research in the late 1970s and early 1980s. It is clear that at least some of them were papers written by others, which Alsabti retyped and submitted to journals listing himself as author. As word spread of his plagiarism, he was eventually driven out of science.[10]

## QUESTIONABLE RESEARCH PRACTICES

While fabrication, falsification, and plagiarism are clearly unethical and cannot be tolerated in science, there are a number of other practices that most scientists consider improper. The topics included in the category of *questionable research practices* are from the National Academy of Sciences report,[11] with five additional topics added.

*Failing to retain significant research data for a reasonable period.* The original data analyzed in a study should be available to others, who may wish to use it in a similar study or who may see a need to check the original analysis to find if it is valid. It is the responsibility of the researcher (or sometimes the laboratory head) to see that the data are kept for sufficient time, just in case. Ideally, data would be kept forever because future generations may want to check a study, but this is not always possible; there simply may be insufficient space to store all data in a laboratory permanently. In some disciplines, central repositories of data have been established and scientists are encouraged to submit copies of their data for general use. Some granting agencies have requirements concerning how long data should be stored. For example, the U.S. National Institutes of Health require that data collected using an NIH grant be retained for three years after the final report is submitted.[12]

*Maintaining inadequate research records, especially for results that are published or are relied on by others.* Much of the work that goes into a scientific study is not included in the published paper. If another researcher needs to know the details of how the research was conducted or what the original data were, then original research notes may need to be consulted. Not keeping records in an understandable and complete form is not proper in science.

*Conferring or requesting authorship on the basis of a specialized service or contribution that is not significantly related to the research reported in the paper.* The list of authors on a paper should indicate those who contributed significantly to the design and conduct of the experiment, analysis and interpretation of the data, and/or the writing of the paper.[13] "Honorary" authorship is sometimes given to individuals not directly involved in the project, but this practice is frowned upon by most scientists. For example, the head of a laboratory may be included even though that person was not involved in the research, or a junior scientist might be included in order to enhance his or her chance for promotion.

Equally improper is leaving someone off of the list of authors who did make a significant contribution. A professor taking sole credit for work done largely by a student fits into this category. So does a senior member of a research team leaving off the name of a junior member who was significantly involved in the study. Of course, there can be much disagreement on the meaning of "significant" with respect to contributions to a piece of research. A paid technician who does most of the routine measurements for a study rarely gets listed as an author. Colleagues who are consulted for advice on how to do the experiment make a contribution to the research, but a decision must be made as to whether the contribution is worthy of authorship. Many scientific papers include an acknowledgments section where minor contributions are mentioned.

*Refusing to give peers reasonable access to unique research materials or data that support published papers.* Scientific progress is based on the free exchange of ideas and information. With few exceptions, research materials and data should be made available to other reputable researchers in the field so that conclusions drawn from the original research can be checked independently and so that the information can be used in additional research projects. The scientist who collects the materials and data, however, should have time to do the original analysis using them. A paleontologist who finds a unique dinosaur fossil, for example, gets to study it first. After his or her description and interpretations are published, though, other paleontologists should be allowed to view the skeleton and confirm or refute the conclusions. It is sometimes difficult for a scientist to painstakingly collect original data and then let others who have not put in the time, money, and sweat use it. But it is for the good of science. The researcher using the data collected by another, of course, must appreciate and credit the work involved in its collection and use the data only for legitimate scientific purposes.

*Using inappropriate statistical or other methods of measurement to enhance the significance of research findings.*    Scientists often have more than one way to analyze data. For example, there may be several statistical techniques that can be used to study the same set of measurements, though one of them is best suited to the particular problem at hand. If the most appropriate method yields results that do not support the hypothesis, it is improper to leave out any presentation and discussion of this analysis and use another method just because the results better support the hypothesis.

*Inadequately supervising research subordinates or exploiting them.*    Professors have a responsibility to see that their students do research properly, in terms of using proper methods and ethical behavior. The same holds true for department and laboratory heads with respect to other professionals they oversee. In addition, professors usually hold some power over their students' futures both in terms of continuing their education and in getting jobs and succeeding in a scientific career. Using that power to take advantage of a student is wrong. Students are not to be treated as research slaves. They should be treated with respect, given credit for the work they do, and educated in the ways of science. Likewise, senior scientists should not take advantage of junior researchers who have less control over their professional destinies.

*Misrepresenting speculations as fact or releasing preliminary research results, especially in the public media, without providing sufficient data to allow peers to judge the validity of the results or to reproduce the experiments.* Scientists sometimes have information of interest to the general public. Communicating with the public is different from communicating with other scientists, though. Other scientists are trained to interpret research results, but most of the public is not. Scientists must make clear when a statement can be backed up with reliable supporting scientific evidence and when it is essentially an unsupported guess, because many in the public do not know how to distinguish between the two. Also, results should not be presented which have not been thoroughly analyzed. In highly competitive fields, it is tempting to release findings to the public as soon as possible in order to get credit for discoveries. Most scientists feel that research results should not be released to the public until they have been through the peer review process. Credit for discoveries usually goes to the first researchers who submit an article to a reputable journal.

In 1989, chemists B. Stanley Pons and Martin Fleischmann held a press conference to announce that they had made a breakthrough in "cold fusion" research. Fusion is the process by which the energy of the sun is generated, and if humans can learn how to control the process, a relatively clean, cheap, and readily available source of energy will be found. Most fusion research involves extremely hot temperatures (hotter than the sun), but fusion at more practical temperatures would be a tremendous advance, hence the name *cold* fusion. Spurred on by Pons's university, which anticipated great recognition and profit, Pons and Fleischmann announced their discovery to the press before their paper had gone through the peer review process. Many other scientists tried to replicate their results, and within a year, many experts were convinced that Pons and Fleischmann had not achieved cold fusion at all. Had Pons and Fleischmann been more careful in experimenting and gotten more feedback from their peers before announcing their results, one of the most controversial and embarrassing events in science in recent history could have been avoided.[14]

Five additional topics were not discussed by the National Academy of Science report but can be added to the list of questionable research practices. These are selective reporting of research, interference, self-plagiarism, misrepresenting one's research record, and conflict of interest.

*Selective reporting of research.*   In a research paper, it is desirable to report that the conclusions are strongly supported by evidence. However, it is wrong to mislead the reader by mentioning only the parts of the study that support the conclusions and leaving out the rest of the work. Improperly deleting data from the analysis, discussed as part of falsification, is a blatant form of selective reporting, but other, more subtle, forms exist. One is analyzing the data in several ways and then only reporting the technique that gives the strongest results. Another is changing the hypothesis after the data are analyzed and reporting the study as though the *post hoc* hypothesis is the one the study was designed to test. Still another form of selective reporting is not mentioning previous studies that do not support the conclusions. Science progresses best when both the strengths and weaknesses of a theory are clearly understood; honesty in reporting is essential.[15]

*Interference.*    It is wrong to harm another scientist's research. A scientist may be tempted to slow or stop the progress of a competitor through intentional damage or theft of property, such as equipment, research notes, and data. Aside from being illegal in most cases, such acts clearly are improper behavior for scientists.[16] It also is improper for a "big name" scientist to take unfair advantage of a less powerful competitor. This could take the form of, for example, attempting to reduce a competitor's access to publication or research money by exerting influence on editors or granting agencies. The advancement of knowledge requires that good research be supported and disseminated solely on its merits, as discussed in the previous chapter under the scientific norm of universalism.

In the 1700s, the British government offered a large monetary prize for the person who could devise an accurate means for determining longitude, which is necessary for ship captains to find their location. While most scientists felt that an astronomical approach would solve the problem, clock maker John Harrison tried to win the prize with a very accurate and durable clock. Nevil Maskelyne was Astronomer Royal, and as such he was a member of the board that evaluated each method and determined the winner of the prize. Maskelyne also was a staunch advocate of the astronomical approach and felt that it was the only method that could work. He thwarted, in a variety of ways, Harrison's attempt to win the prize. Harrison devoted more than forty years of his life to the longitude problem and met all of the requirements to win the prize, but he did not, largely due to Maskelyne's interference.[17]

Another form of interference is improper allegations of misconduct. Being charged with scientific misconduct can seriously damage a career, even if the charges are officially found to be false. Allegations should be made on scholarly grounds only and not for any malicious or vindictive reasons or to cause trouble for someone whose research results are disagreeable to some person or group.

*Self-plagiarism.*   Because scientists often are judged by the number of publications they produce, there is a temptation to publish the same study in more than one journal. However, it is generally considered improper to do so. It is acceptable to include previous research results in a new paper if additional work is also presented. The scientist should, however, cite the previous work even if it is his or her own paper so as not to mislead the reader. The practice of publishing research results in conference proceedings, which often are not reviewed, and then publishing the same work in reviewed journals is common, though discouraged by some. Publishing the same work in two refereed journals is clearly improper.[18]

*Misrepresentation of one's research record.*   Scientists keep a *curriculum vita*, the academic version of a business resume, which lists positions held, authored or coauthored publications and presentations, research grant proposals funded, and other pertinent information. The contents of the vita are used to judge scientists for jobs, promotions, and research grants. Some scientists make themselves look more impressive by listing publications that do not exist and talks that were never given.[19] Such practices are fraudulent. Duplicate publications, discussed previously, are another means of "padding the vita." A related problem is the improper listing of unpublished work. A paper or book "in progress" is currently being written, "submitted for publication" means it has been sent to a journal editor or publisher and is currently under consideration, and "in press" means accepted for publication but not yet published. A paper listed as "in press" is more impressive than one "submitted for publication" because papers in the latter category might get rejected; similarly, works "in progress" may never get finished. To lie in one's vita about where a paper or book is along the trail to publication is improper.

*Conflict of interest.*   Certain situations arise in which a scientist's results may be subjected to extra scrutiny because of potential conflicts of interest. As Roger Porter states, "Conflicts of interest have been variously defined, but the best summary is the oldest: 'No one can serve two masters, for either he will hate the one and love the other, or he will be devoted to one and despise the other. You cannot serve both God and mammon' (Matthew 6:24). A person with a conflict of interest has two masters to serve and these two masters do not always have coincident needs."[20] In other words, if a scientist is concerned with both the advancement of knowledge and the making of money (or its biblical version, *mammon*) a situation *may* arise where one of them must be compromised. Examples of conflicts of interest include research on the environmental effects of a pesticide paid for by the company manufacturing the pesticide and research on the effectiveness of a drug done by a scientist who has invested money in the company selling the drug. Keep in mind that such conflicts of interest do not mean that the work of the scientist is invalid or untrustworthy, but the possibility exists that it is. Many scientists are even more skeptical than normal when evaluating such work. It is important that the researcher state clearly that such *potential* conflicts exist. To hide that information is improper. Of course, committing scientific fraud for personal economic gain or to enhance a political position is wrong.

Most discussion of conflict of interest involves the possibility that research done by a scientist may be biased for financial gain. The U.S. Public Health Service (PHS), for example, has proposed rules which ". . . require disclosure of 'significant financial interests' of the Investigator that would reasonably appear to be directly and significantly affected by the research funded by PHS. . . ."[21] Significant financial interests are interpreted as the scientist having the potential of profiting from a company which may benefit from the research or a company competing with one which may profit from the work. The U.S. National Science Foundation has similar rules.[22]

Another area of conflict of interest is the misuse of proprietary information. If, for example, a scientist does a peer review of a paper for a journal and learns from the paper that a new drug is significantly more effective than competing drugs, the reviewer should not invest money in the company selling the new drug before the paper is published.

An example of possible conflict of interest involves a test of a drug used on eyes by researchers who also formed and owned stock in the company planning to market the drug. As Booth describes it, ". . . it may turn out to be a tale of greed or one of naïveté, a morality play about what happens to researchers with stock options or a cautionary tale about the dangers of careless enthusiasm. Either way, the story takes place in an ethical no-man's land where the interests of academic science and business collide."* Briefly put, the initial research on the drug done by the scientists with a financial stake in the outcome suggested that it was a promising treatment. There was concern over both the potential conflict of interest and the methods used in the study, which prompted a second study, conducted by scientists with no personal financial ties to the company. This second study suggested that the drug was no more effective than a placebo.[23]

A topic related to conflict of interest is *conflict of conscience*. A scientist employed by an environmental organization with a strong stand against pesticide use who publishes research on the environmental effects of pesticides should make clear the political agenda of the sponsoring organization. Another example of conflict of conscience is a researcher who is morally opposed to the use of animals in research who is asked to do a peer review of a study using animals.[24] The scientist's views should have no effect on the review of the scientific merits of the study, and he or she should either decline to do the review or state clearly that such views might affect the objectivity of the evaluation.

## RESEARCH WITH HUMAN AND ANIMAL SUBJECTS

Research done with human or animal subjects raises ethical issues beyond those already discussed, and there are constraints on how such research can be conducted. The potential benefits of the research to knowledge must be weighed against the potential for temporary pain and discomfort and the possibility of permanent injury to the subjects. Each institution where human or animal subjects are used must have review boards to ap-

---

*Reprinted with permission from William Booth, "Conflict of Interest Eyed at Harvard," *Science* (1988) 242:1497. Copyright © 1988 American Association for the Advancement of Science.

prove each proposed research project. The boards are composed of members who are not involved in the projects.

The most important ethical requirement in research using human subjects is *informed consent*. Quoting from the World Medical Association Declaration of Helsinki:[25]

> In any research on human beings, each potential subject must be adequately informed of the aims, methods, anticipated benefits and potential hazards of the study and the discomfort it may entail. He or she should be informed that he or she is at liberty to abstain from participation in the study and that he or she is free to withdraw his or her consent to participation at any time. The physician should then obtain the subject's freely-informed consent, preferably in writing.

In informed consent, the subject must be able to understand the potential consequences of the study on his or her physical and mental well-being and agree to participate without any coercion. In the case of children or others unable to make the decision for themselves, parents or legal guardians must decide if consent is to be given.[26]

In some psychological and social research using human subjects, the concept of informed consent is modified. If a subject is told that she or he is being observed for a study on racist attitudes, for example, that knowledge would likely affect his or her behavior, and the results of the research would be misleading. Deception in consent is allowed only if the risk to the subject is minimal or nonexistent, the rights and welfare of the subject are not affected, the deception is scientifically necessary (the study results would be incorrect without it), and if possible, the true nature of the research is explained after the subject is finished participating. Keep in mind, though, that there is much debate over the extent to which deception in consent should be allowed.[27]

An example of unethical research using human subjects is the Tuskegee Syphilis Study, begun in 1932. A group of 399 African-American men with syphilis and another 201 without the disease were studied for decades to determine how untreated syphilis affects the body. The nature of the study was never explained to the men, who were mostly poor and illiterate. Even as more effective medical treatment became available, the men were not treated, and the syphilis eventually killed many of them. Informed consent, as currently defined, was missing from the Tuskegee Study and the potential for an increase in understanding of the effects of syphilis did not justify the premature deaths of some of the subjects.[28]

In the case of animal subjects used in research, proposed projects must be approved by institutional review boards, as previously discussed. In addition, there are rules for the care of the animals, such as housing requirements, veterinary oversight, and how animals are to be obtained. There is a heated debate concerning the ethical issue of whether scientists should use animals in research at all. The moral dimensions of the debate are beyond the scope of this book,[29] but from a practical point of view, animal research has greatly improved human and animal health and the general scientific understanding of biological phenomena.[30]

There is debate as to the implications of using results from unethically conducted research. An extreme example of this problem is a study of the effects of cold water on humans conducted on Nazi concentration camp victims during World War II. Hundreds of men were submerged in cold water for extended periods of time and monitored to determine the most effective means of warming bodies after exposure and also to determine the body temperature at which death occurs. Many of the subjects died during the tests, which apparently was the intention of the researchers.[31] While acknowledging the horrendous nature of these experiments, the question arises as to whether or not the results should be used by subsequent researchers on the subject. On the one hand, the victims' vile deaths might have provided information that could be used to save other lives and thus have provided some benefit to humankind. On the other hand, use of the concentration camp research results might be interpreted as condoning unethical research and thus indirectly encouraging it. The *New England Journal of Medicine,* for example, has taken the latter position and states that it will not publish the results of any unethically conducted studies.[32] Such a stance extends the scientific norm of universalism, discussed in Chapter 4. Universalism is the idea that scientists are to be judged only by their competence as scientists and such judging should include not only the competence of the scientist in conducting rigorous research, but ethical research as well.

### WHISTLEBLOWING

All scientists have a responsibility to minimize the existence of scientific misconduct. If a scientist knows with certainty that unethical behavior has occurred, he or she has an obligation to report such information to the community of scientists. Unfortunately, such reporting, or *whistleblowing*, is very difficult for all concerned.[33] Whistleblowers are often subjected to professional attacks by individuals who are threatened by the accusations. This can include the accused scientist, as well as those who fear the bad publicity for a laboratory, university, or funding agency. For example, Robert Sprague, who blew the whistle on Stephen Bruening, as discussed earlier in this chapter, found himself under investigation and subjected to what many would consider professional attacks.[34] If serious misconduct occurs, it is the ethical responsibility of scientists to blow the whistle. However, it should be done very cautiously and carefully, and junior scientists or students should seek the support of senior scientists before proceeding with the reporting. Whistleblowing itself is unethical when done for reasons other than matters of research integrity; it is not to be used to make trouble for personal enemies.[35]

Proper ethical practice is as important as any other aspect of scientific research. In fact, scientific misconduct can turn an otherwise brilliant piece of research into one that harms the quest for knowledge. Scientists have a moral obligation to not let this happen.

### PROBLEMS

1. Why is integrity so important for scientists?

2. What is the difference between fabrication and falsification? Give an example of each.

3. "Plagiarism is more than copying another's words." Discuss this statement.

4. Why are research records so important in science? Why must others be allowed to see them?

5. Why should we care if the author list on a scientific paper does not reflect who did the work on the study?

6. What is wrong with skipping the peer review process and disseminating scientific results directly to the public?

7. Does a conflict of interest mean that the research is not to be trusted? Discuss.

8. Give your opinion (and justify it) on the use of animals in scientific research.

9. From an ethical point of view, why is informed consent important in research using human subjects.

## NOTES

1. LaFollette, Marcel C., 1992, *Stealing into Print: Fraud, Plagiarism, and Misconduct in Scientific Publishing*, Berkeley: University of California Press, p. 1.

2. Maddox, John, 1995, "Restoring Good Manners in Research," *Nature* 376: 113.

3. Mohr, Hans, 1979, "The Ethics of Science," *Interdisciplinary Science Reviews* 4: 48.

4. See, for example, Bell, Robert, 1992, *Impure Science: Fraud, Compromise and Political Influence in Scientific Research*, New York: John Wiley & Sons, Inc., p. 105–43.

5. Committee on Science, Engineering, and Public Policy (U.S.). Panel on Scientific Responsibility and the Conduct of Research, 1992, *Responsible Science: Ensuring the Integrity of the Research Process*, vol. I, Washington, D.C.: National Academy Press, p. 5 6. Discussion of ethical issues not covered in this chapter, such as research that causes environmental destruction and racial and gender bias in research, can be found in Shrader-Frechette, Kristin, 1994, *Ethics of Scientific Research*, Lanham, Maryland: Rowman and Littlefield Publishers, Inc.

6. Bell, 1992, p. 105–11; Robert Sprague, personal communication, 29 July 1994.

7. Webster, Gary D.; Rexroad, Carl B.; and Talent, John A.; 1993, "An Evaluation of the V. J. Gupta Conodont Papers," *Journal of Paleontology* 67: 486 93. See also Talent, John A., 1989, "The Case of the Peripatetic Fossils," *Nature* 338: 613–15.

8. Holton, Gerald, 1978, "Subelectrons, Presuppositions, and the Millikan-Ehrenhaft Dispute," in *Historical Studies in the Physical Sciences*, vol. 9, eds. Russell McCormmach and Lewis Pyenson, Baltimore: The Johns Hopkins University Press, p. 161–224. (It is debatable whether or not Millikan's falsification was improper given the practices of science early in the twentieth century, but there is no doubt that such behavior is unacceptable now.)

9. Committee on the Conduct of Science, National Academy of Sciences, 1989, *On Being a Scientist*, Washington, D.C.: National Academy Press, p. 6.

10. Broad, William; and Wade, Nicholas, 1982, *Betrayers of the Truth*, New York: Simon & Schuster, p. 203–11. See also Miller, David J., 1992, "Plagiarism: The Case of

Elias A. K. Alsabti," in *Research Fraud in the Behavioral and Biomedical Sciences,* eds. David J. Miller and Michel Hersen, New York: John Wiley and Sons, p. 80–96.

11. Committee on Science, Engineering, and Public Policy, 1992, p. 28.

12. Macrina, Francis L., 1995, *Scientific Integrity: An Introductory Text with Cases,* Washington, D.C.: ASM Press, p. 47.

13. Sigma Xi, 1991, *Honor in Science,* Research Triangle Park: Sigma Xi, The Scientific Research Society, Inc., p. 23–27.

14. Huizenga, John R., 1992, *Cold Fusion: The Scientific Fiasco of the Century,* Rochester: University of Rochester Press. See also Close, Frank, 1991, *Too Hot to Handle: The Race for Cold Fusion,* Princeton: Princeton University Press.

15. Bailar, John C., 1995, "The Real Threats to the Integrity of Science," *The Chronicle of Higher Education,* 41, 21 April 1995, p. B1.

16. Commission on Research Integrity (Kenneth J. Ryan, chair), 1995, "Integrity and Misconduct in Research," U.S. Department of Health and Human Services, Public Health Service, p. 11–13.

17. Sobel, Dava, 1995, *Longitude: The True Story of a Lone Genius Who Solved the Greatest Scientific Problem of His Time,* New York: Walker and Company. While Harrison did receive much of the prize money for his invention, he never was given the honor of winning the Longitude prize itself.

18. Kohn, Alexander, 1986, *False Prophets: Fraud and Error in Science and Medicine,* Oxford: Basil Blackwell, p. 150–52.

19. Sekas, Gail, and Hutson, William R., 1995, "Misrepresentation of Academic Accomplishments by Applicants for Gastroenterology Fellowships," *Annals of Internal Medicine* 123: 38–41.

20. Porter, Roger J., 1992, "Conflicts of Interest in Research: The Fundamentals," in *Biomedical Research: Collaboration and Conflict of Interest,* eds. Roger J. Porter and Thomas E. Malone, Baltimore: The Johns Hopkins University Press, p. 121–34; quote is on p. 125.

21. Department of Health and Human Services, Public Health Service, 1994, "Objectivity in Research," *Federal Register* 59, no. 123, 28 June 1994, p. 33242–51; quote is on p. 33244.

22. National Science Foundation, 1994, "Investigator Financial Disclosure Policy," *Federal Register* 59, no. 123, 28 June 1994, p. 33308–12.

23. Booth, William, 1988, "Conflict of Interest Eyed at Harvard," *Science* 242: p. 1497–99; quote is on p. 1497.

24. Macrina, 1995, p. 178–79.

25. Ibid., p. 158.

26. Ibid., p. 137–51. *See also* Department of Health and Human Services, National Institutes of Health, Office of Protection from Research Risks, 1991, *Protection of Human Subjects,* Code of Federal Regulations, Title 45, Part 46.

27. Penslar, Robin Levin, 1995, *Research Ethics: Cases and Materials,* Bloomington: Indiana University Press, p. 112–13. See also Sieber, Joan E., 1984, "Informed Consent and Deception," in *NIH Readings on the Protection of Human Subjects in Behavioral and Social Science Research: Conference Proceedings and Background Papers,* ed. Joan Seiber, Frederick,

Maryland: University Publications of America, p. 39–77. See also Shrader-Frechette, 1994, p. 7–8.

28. Jones, James H., 1993, *Bad Blood: The Tuskegee Syphilis Experiment (New and Expanded Edition)*, New York: The Free Press.

29. For useful introductions to the ethical dimensions of the animal rights debate, see Macrina, 1995, p. 97–136 and Smith, Jane A., and Boyd, Kenneth M., eds., 1991, *Lives in the Balance: The Ethics of Using Animals in Biomedical Research*, Oxford: Oxford University Press. See also the series of articles in the February 1997 issue of *Scientific American* 276, no. 2: 79–93. For an informative discussion of the radical element of the animal rights movement, see Lutherer, Lorenz Otto, and Simon, Margaret Sheffield, 1992, *Targeted: The Anatomy of an Animal Rights Attack*, Norman: University of Oklahoma Press.

30. Committee on the Use of Animals in Research, 1991, *Science, Medicine and Animals*, Washington, D.C.: National Academy Press. See also Botting, Jack H., and Morrison, Adrian R., 1997, "Animal Research Is Vital to Medicine," *Scientific American* 276, no. 2: 83–85.

31. Berger, Robert L., 1990, Nazi Science—The Dachau Hypothermia Experiments, *New England Journal of Medicine* 322: 1435–40. Berger concludes that not only were the experimental methods viciously unethical, but the experimental rigor was so poor that, even though the results had been used in subsequent studies, such use is unwarranted on purely scientific grounds as well. Berger mentions another disgusting study done to improve knowledge on blood clotting in wounded soldiers. These experiments involved simulating combat wounds by amputating arms and legs of unanesthetized inmates or shooting the prisoners in the chest and neck.

32. Angell, Marcia, 1990, "The Nazi Hypothermia Experiments and Unethical Research Today," *New England Journal of Medicine* 322: 1462–64.

33. Sigma Xi, 1991, p. 29–32

34. Bell, 1992, p. 108–11.

35. Commission on Research Integrity, 1995, p. 21–24. See also Gunsalus, C. K., 1998, "How to Blow the Whistle and Still Have a Career Afterwards," *Science and Engineering Ethics* 4: 51–64.

# CHAPTER 6

# CRITICAL THINKING
# AND SCIENCE

Muddleheadedness has always been the sovereign force in human affairs—a force far more potent than malevolence or nobility. It lubricates our hurtful impulses and ties our best intentions in knots. It blunts our wisdom, misdirects our compassion, clouds whatever insights into the human condition we manage to acquire. It is the chief artisan of the unintended consequences that constitute human history.

*Paul R. Gross and Norman Levitt, 1994*[1]

Science, like all human endeavors, suffers when "muddleheadedness" prevails. The ability to think critically is crucial for scientists. Scientists must be able to make decisions based primarily on reason, not wholly on emotion, and to think fairly, creatively, and constructively about both their own work and the work of others. *Critical thinking* is an acquired skill and one on which all scientists have to work throughout their careers.

Critical thinking is a rational response to questions that cannot be answered definitively and for which all the relevant information may not be available. *It is . . . an investigation whose purpose is to explore a situation, phenomenon, question, or problem to arrive at a hypothesis or conclusion about it that integrates all available information and that can therefore be convincingly justified.* In critical thinking, all assumptions are open to question, divergent views are aggressively sought, and the inquiry is not biased in favor of a particular outcome.

Joanna Kurfiss, 1988[3]

This chapter introduces critical thinking as it applies to science, but keep in mind that critical thinking is in no way limited to scientific pursuits. It is a skill that all people should develop personally and encourage in others. In the words of Brooke Moore and Richard Parker, "Critical thinking is simply the careful, deliberate determination of whether we should accept, reject, or suspend judgment about a claim—and of the degree of confidence with which we accept or reject it."[2]

---

One of the most difficult operations for imperfect mortals is the making of distinctions, of stopping opinion and belief part way, of accepting qualified ideas. But the individual who can modify or correct beliefs molded by personal interest or the influence of his rearing is rare. . . . It is easy to be wise in retrospect, uncommonly difficult in the event.

Wallace Stegner, 1962[4]

---

As opposed to critical thinking, where thought processes are carefully controlled, *imaginative thinking* is when the mind is allowed to wander freely, as in a daydream. Critical thinking is used most of the time in science, but imaginative thinking is important in coming up with new ideas, as in hypothesis generation or in figuring out new ways to solve a problem encountered in research. Both play important roles in science. Imaginative thinking was discussed briefly in Chapter 3 with respect to hypothesis generation.

---

The usual procedure in research is to follow critical thinking and only when that fails to switch to imaginative thinking to try to find a way around the difficulty. . . . Then if some promising idea comes up one reverts to critical thinking to work it out.

W. I. B. Beveridge, 1980[5]

---

## CRITICAL THINKING STRATEGIES

Richard Paul has developed a list of thirty-five strategies of critical thinking, divided into two categories: affective strategies and cognitive strategies.[6] *Affective strategies* involve attitudes and behavior related to critical thinking. *Cognitive strategies* involve the basic skills involved in critical thinking and the processes by which these skills are used. Each of Paul's strategies will be listed (in italics) and briefly discussed in terms of applications to science.

## Affective Strategies

*1. Thinking independently.*    Scientists should not accept theories or other information without determining for themselves if they are reasonable. The opinions of others are important, but the final decision to accept or reject a scientific statement should come from the individual scientist. Unfortunately, independent thought is difficult for many people and is an acquired skill.

---

The human foot was not built for ballet. Only with discipline, training, and pain, can it endure the strain and produce beauty. The human mind was not built for independent thinking. Only with discipline, training, and pain, can it endure the strain and produce knowledge.

William Daly, 1995[7]

---

*2. Developing insight into egocentricity or sociocentricity.*    A scientist should think carefully about biases and how they affect his or her work. In other words, a scientist should seek to understand the differences between his or her perceptions of something and its objective reality and try to minimize the difference. The same can be said for the biases found in groups, both scientific groups (physicists as a group view the world somewhat differently than chemists as a group, for instance) and cultural groups (the French may see the world differently than the British, stereotypically speaking).

*3. Exercising fair-mindedness.*    No matter how strongly a scientist feels about a theory, for example, the possibility of being wrong is always present, and he or she must carefully consider the strengths and weaknesses of opposing viewpoints. Scientists need to make sure that they support a theory because of logic and evidence and not for other reasons.

But that is the way of the scientist. He will spend thirty years in building up a mountain range of facts with the intent to prove a certain theory; then he is so happy in his achievement that as a rule he overlooks the main chief fact of all—that his accumulation proves an entirely different thing. When you point out this miscarriage to him he does not answer your letters; when you call to convince him, the servant prevaricates and you do not get in. Scientists have odious manners, except when you prop up their theory; then you can borrow money of them.

Mark Twain, 1917[8]

*4. Exploring thoughts underlying feelings and feelings underlying thoughts.* A popular image of a scientist is that of an emotion-free entity, but of course, this is not the case in reality. In pursuing their research, scientists experience joy, anger, depression, and most other emotions, and these affect the scientist's work. Scientists should see that reason and emotion are interconnected and understand how this relationship affects their thinking and actions.

The inner nature of science within the scientist is both emotional and intellectual. The emotional element must not be overlooked, for without it there is no sound research on however odd and dull-seeming a subject. As is true of all of us, an emotion shapes and informs the scientist's life; at the same time an intellectual discipline molds his thinking, stamping him with a character as marked as a seaman's although much less widely understood.

Oliver La Farge, 1942[9]

*5. Developing intellectual humility and suspending judgment.* Scientists need to acknowledge the limitations of their understanding of something. As Paul states, this ". . . does not imply spinelessness or submissiveness. It implies the lack of intellectual pretentiousness, arrogance, or conceit. It implies insight into the foundations of one's beliefs: knowing what evidence one has, how one has come to believe, what further evidence one might look for or examine."[10] Scientists should feel at ease saying "I don't know," when such a response is appropriate. They should also qualify their statements on scientific matters to acknowledge what is not known about the subject.

> As a human being, one has been endowed with just enough intelligence to be able to see clearly how utterly inadequate that intelligence is when confronted with what exists. If such humility could be conveyed to everybody, the world of human activities would be more appealing.
>
> Albert Einstein, 1932[11]

*6. Developing intellectual courage.*   Scientists should deal with ideas directly and honestly and that includes confronting unreasonable aspects of popular ideas and acknowledging reasonable parts of unpopular ideas. Nonconformity, when justified, is a sign of courageous critical thought, even though the consequences of nonconformity sometimes are severe.

*7. Developing intellectual good faith or integrity.*   What Paul states in general applies to the specific case of scientists: "Critical thinkers recognize the need to be true to their own thought, to be consistent in the intellectual standards they apply, to hold themselves to the same rigorous standards of evidence and proof to which they uphold others, and to honestly admit discrepancies and inconsistencies in their own thought and action."[12]

> [Science] warns me to be careful how I adopt a view which jumps with my preconceptions, and to require stronger evidence for such belief than for one to which I was previously hostile.
>
> My business is to teach my aspirations to conform themselves to fact, not to try and make facts harmonise with my aspirations.
>
> Thomas Henry Huxley, 1860[13]

*8. Developing intellectual perseverance.*   Critical thought is not easy. Before making a judgment on an idea, the scientist should take the time and make the effort to be sure that the idea has been analyzed fully and carefully.

*9. Developing confidence in reason.*   The careful use of reason leads to the advancement of science and is far superior to the acceptance of ideas based on emotional pleas, one-sided arguments, or governmental force.

---

It must be remembered that logical, skeptical thought is something the child must acquire, that magical thinking comes to us much more naturally. We gradually learn to trust logic and rationality because our experience teaches us that they serve us well. Yet, lurking not so far beneath the surface is that magical mode of thinking which came so automatically to us in childhood.

James Alcock, 1991[14]

---

*Cognitive Strategies*

*10. Refining generalizations and avoiding oversimplifications.*   The subject matter of science is often highly complex, and in order to make sense of it, scientists deal with reality in a simplified form. However, a scientist must be able to simplify without losing track of the whole situation. A good model, whether verbal, physical, or mathematical, keeps the essence of a system while simplifying it into a manageable form.

*11. Comparing analogous situations: transferring insights to new contexts.*   Scientists should look for novel ways to use ideas. They should consider whether a new theory or research method in one discipline could be useful in another.

*12. Developing one's perspective: creating or exploring beliefs, arguments, or theories.*   Most scientists approach their subject from a particular perspective, perhaps as simulation modelers or laboratory experimenters, or from a given philosophical view. They must be keenly aware, however, of the strengths and limitations of both their perspective and alternative approaches. Scientists should acknowledge that there are many approaches to the acquisition of knowledge, and one's own is not necessarily better than another's.

*13. Clarifying issues, conclusions, or beliefs.* Before evaluating an issue and drawing a conclusion about it, scientists must make sure that the idea is presented clearly and completely. They must understand what information is required before a conclusion can be reached.

*14. Clarifying and analyzing the meanings of words or phrases.* In analyzing an idea, the specific meanings of words and phrases must be clearly understood. Otherwise, differing interpretations of language may lead to unnecessary disagreements over a concept. In the words of William Gray, "[a] verbal dispute occurs when two or more people think that they are arguing about objects named by their words when, in fact, they are really arguing about the words themselves."[15]

*15. Developing criteria for evaluation: clarifying values and standards.* It is important that the information required to make a decision be explicitly stated and to understand how different points of view might affect how an idea is evaluated.

*16. Evaluating the credibility of sources of information.* Conclusions should be drawn from reliable information, and scientists need to learn to discern the reliability of the information sources. For example, when the methods used in an experiment are not clearly stated, the results should not be considered reliable. In addition, a scientist should avoid the temptation to use unreliable information just because it supports her or his conclusions. The fallibility of memory and eyewitness accounts of events should be appreciated and is discussed in Chapter 7.

*17. Questioning deeply: raising and pursuing root or significant questions.* A scientist should always keep a broad perspective in mind when conducting research or learning about a topic. Specific issues should be evaluated in terms of what they contribute to the "big picture."

*18. Analyzing or evaluating arguments, interpretations, beliefs, or theories.* Scientists, like all ". . . critical thinkers recognize the importance of asking for reasons and considering alternative views. They are especially sensitive to possible strengths of arguments that they disagree with, recognizing the tendency of humans to ignore, oversimplify, distort, or otherwise unfairly dismiss them."[16]

*19. Generating or assessing solutions.* In finding a solution to a problem, or evaluating someone else's solution, scientists should ". . . formulate problems clearly, accurately, and fairly, rather than offering a sloppy, half-baked description and then immediately leaping to solutions."[17] They should also actively seek alternative solutions to see if one provides a better explanation of the phenomena under study.

*20. Analyzing or evaluating actions and policies.* Scientists sometimes need to evaluate the behavior of others (as in breeches of ethical standards by a researcher) and of groups (as in national science policy or university policy toward research). Such actions and policies should be evaluated in terms of carefully thought-out standards, not impulsive reactions.

*21. Reading critically: clarifying or critiquing texts.* Scientists should be skeptical and careful readers. When reading scientific literature, a scientist must make sure that he or she understands what is said and then must decide whether to agree, disagree, or not pass judgment on the conclusions. The scientist should also attempt to determine what information or viewpoints have not been included in the paper and how they might affect his or her judgment on the conclusions.

*22. Listening critically: the art of silent dialog.* As with reading, scientists should be careful listeners, attempting not only to understand what is being directly said by the speaker but also to appreciate the speaker's point of view and how that affects what is said. Unlike reading, listening often allows one to ask questions to clarify misunderstood points quickly.

*23. Making interdisciplinary connections.* Paul states, "Critical thinkers do not allow the somewhat arbitrary distinctions between academic subjects to control their thinking. When considering issues which transcend subjects, they bring relevant concepts, knowledge, and insights from many subjects to the analysis."[18]

*24. Practicing Socratic discussion: clarifying and questioning beliefs, theories, or perspectives.*   Asking good questions is a skill that all scientists should acquire. Questions can help the listener better understand what the speaker or writer is saying and help the speaker better develop her or his ideas. Critical thinkers welcome good questions from others.

*25. Reasoning dialogically: comparing perspectives, interpretations, or theories.*   If two scientists disagree on something, they should constructively discuss their differences. Even if they do not come to an agreement, the dialog will help each better understand the situation. Such a dialog can also be done by an individual scientist seeking to understand differing perspectives on a problem. Bertrand Russell writes: "For those who have enough psychological imagination, it is a good plan to imagine an argument with a person having a different bias. . . . I have sometimes been led actually to change my mind as a result of this kind of imaginary dialogue, and, short of this, I have frequently found myself growing less dogmatic and cocksure through realizing the possible reasonableness of a hypothetical opponent."[19]

*26. Reasoning dialectically: evaluating perspectives, interpretations, or theories.*   A form of dialogical thinking is dialectical thinking in which conflicting ideas are set against each other in order to understand the strengths and weaknesses of each, as in a formal debate.

*27. Comparing and contrasting ideals with actual practice.*   Many of the ideal situations presented in scientific textbooks are generalizations that are difficult to find in specific situations in either nature or society. Scientists should always be aware of this distinction and work to minimize the difference between ideal and actual.

*28. Thinking precisely about thinking: using critical vocabulary.*   "One possible definition of critical thinking is the art of thinking about your thinking while you are thinking in order to make your thinking better: more clear, more accurate, more fair."[20] This quote from Paul makes sense, if you think about it. Paul also encourages the use of analytical vocabulary when thinking about thinking, ". . . such terms as 'assume,' 'infer,' 'conclude,' 'criteria,' 'point of view,' 'relevance,' 'issue,' 'elaborate,' 'ambiguous,' 'objection,' 'support,' 'bias,' 'justify,' 'perspective,' 'contradiction,' 'consistent,' 'credibility,' 'evidence,' 'interpret,' 'distinguish.' "[21]

*29. Noting significant similarities and differences.*   It is important to appreciate the difference between superficial and substantial similarities and differences. It is also important to understand that two things that are different from one point of view may be similar from another. For example, political affiliations may be important to a political scientist but have no significance to a medical researcher.

*30. Examining or evaluating assumptions.*   In a scientific discussion, some assumptions are stated and others are not. Before judging an idea, critical thinkers try to determine all assumptions important to a theory and determine if they are valid.

*31. Distinguishing relevant from irrelevant facts.*   The facts used to support or detract from a scientific idea should be evaluated to determine if they are pertinent to the discussion. Irrelevant information detracts from the drawing of reasonable conclusions.

*32. Making plausible inferences, predictions, or interpretations.*   The facts and concepts used to support or detract from a scientific idea should be judged for their implications in that particular case. The results of an experiment, for example, should be evaluated in terms of other possible inferences that could be drawn, perhaps by scientists with other points of view.

*33. Evaluating evidence and alleged facts.*   Quoting Paul again: "Not everything offered as evidence should be accepted. Evidence and factual claims should be scrutinized and evaluated. Evidence can be complete or incomplete, acceptable, questionable, or false."[22]

*34. Recognizing contradictions.*    Scientists should actively seek to find the real differences in opposing viewpoints and concentrate on studying these differences in an attempt to resolve the conflict. They should also seek to be consistent in the way judgments are made, especially with respect to ideas with which they tend to agree and ideas with which they tend to disagree.

*35. Exploring implications and consequences.*    When a scientist decides to accept a scientific statement as valid, he or she should also examine how that statement can be used in science. By studying the implications and consequences of an idea, a more complete understanding of the concept is achieved.

Any listing of critical thinking strategies is somewhat arbitrary, and the reader may have thought of other ways of organizing the list and other strategies that might be included (especially if this chapter was read critically). Paul's list is used here because it is detailed and comprehensive, though not exhaustive. The following cognitive strategies, for example, can be added to Paul's list.

*Judging ideas and actions in historical context.*    Scientists of the past are not less intelligent than scientists today, they simply had a smaller base of knowledge on which to build. Now we take the concept of gravity for granted, but scientists before Newton cannot be expected to have understood gravity and incorporated it into their theories. Before the 1960s, many geologists argued against the theory of continental drift, which is generally accepted today. Most of their arguments against continental drift were quite valid, based on what was then known about geophysics.

*Seeking primary sources of information.*    Scientists should use original sources of information, whenever possible. They should not rely on textbooks, summaries, or discussions by others (known as *secondary sources*) to understand a theory. When an idea is presented by someone other than the originator, typically, some information is lost or changed. (The book you are reading, for example, is a secondary source and only serves as an introduction to the ideas presented; if the ideas are important in a scientist's research, the primary sources should be sought.) In science, most primary sources are research papers or books. Primary sources are not always available or practical, though. Most physicists are not expected to read Newton's *Principia* in its original Latin, for example, though they should think about the possibility of missing some of the subtleties of his ideas when relying on translations or textbook interpretations.

This list of skills should be read carefully by scientists in order to be introduced to critical thinking and then reviewed occasionally to remind them to consciously practice the skills and behaviors involved. Strategy 27, concerning the difference between ideal and actual practice, applies to scientists and critical thinking. Ideally, all scientists are superb critical thinkers. They have, ideally, an impeccable ability to argue logically, the subject of the next section.

## COMMON FALLACIES

Another aspect of critical thinking is the ability to recognize errors in arguments. In science, as in other areas of knowledge, when a claim is made, reasons must be given for accepting it. However, those reasons must be logically appropriate. Where the reason or reasons given in support of a claim are inadequate for accepting the claim, then a *fallacy* has been committed. Presented here are some common fallacies relevant to science and such related topics as pseudoscience, the subject of the next two chapters. Just as the list of critical thinking skills presented previously was modified from the work of Richard Paul, this list of common fallacies has been taken from more detailed discussions of the topic by William Gray and by Theodore Schick and Lewis Vaughn and modified to fit the topics of this book.[23] Each of the fallacies will be described, and one or two hypothetical examples will be given.

For this discussion, an argument is one or more reasons, or *premises*, given to support a claim, or *conclusion*. If the premises do not adequately support the conclusion, then a fallacy has been committed, and the argument does not justify accepting the conclusion. Keep in mind, though, that just because an argument is fallacious does not mean that the conclusion is necessarily false. For example, "*The Scientific Endeavor* is one of the finest books ever written because the author is such a wonderful person" is a fallacious statement. The premise (assume, for the sake of discussion, that the author really is a wonderful person) simply does not give sufficient reason to accept the conclusion; most wonderful people do not write great books. However, that does not mean that the conclusion or the premise are necessarily wrong, it is just that no good reason has been given to accept them. Before accepting or rejecting the conclusion, more information must be gathered. Fallacies can be grouped into three categories: unacceptable premises, irrelevant premises, and insufficient premises.

*Unacceptable premises* involve problems with the reasons given for accepting the claim. *Begging the question*, also known as "circular reasoning," is a fallacy in which the conclusion is the same as the reasons for accepting it. "The Bible is correct about the origins of humans because the Bible is always right" is an example of begging the question. "To test the theory that force equals mass times acceleration, I measured mass and acceleration and then multiplied the two values to determine force. The agreement was perfect; therefore, the theory is correct." In this hypothetical example, the circular reasoning should be clear, in that the theory determined the results of its test. In more realistic and complex scientific situations, however, the fact that the results are determined by the hypothesis being tested may not be so obvious.

*False dilemma* is another fallacy with unacceptable premises. Here the argument is presented as though there are only two competing explanations, when there are more than two. "Either I become a professional scientist or I'm a failure in life" is fallacious because there are many ways to be successful in life without being a scientist. "If I don't show that the theory is correct, then I have wasted my time" is another example.

*Insufficient premises* provide no, or at best only partial, support for accepting the conclusion. *Hasty generalization* is a fallacy in which a conclusion is drawn about an entire group from an unrepresentative sample of the group. "I had lunch with three sociologists yesterday and that taught me that sociologists don't know anything about psychology" and "All butterflies behave like the ones I watch in my backyard" are examples of hasty generalization.

The fallacy of *faulty analogy* involves claims that just because things have similar characteristics they must be similar in other ways as well. "Kilauea and Fuji are both active volcanoes, and Kilauea erupts most years, so Mt. Fuji must erupt yearly, too" is a faulty analogy because they are different types of volcanoes with different eruption patterns.

*False cause* is another fallacy involving insufficient premises. When one event typically follows another, that does not necessarily mean that the first caused the second. U.S. presidential elections, for example, usually occur several months after the Summer Olympics, but that does not mean that the Olympics cause the elections. Assuming a cause only because of a correlation is fallacious reasoning.

The remaining fallacies fit into the category of *irrelevant premises*, where the reasons given to support a claim are unrelated to the claim. An example is *ad hominem* attacks, which occur when an idea is discounted because of the person who presented it. (*Ad hominem* can be loosely translated from Latin as "attack the person.") In science, an idea must be judged on its merits, and it does not matter if the person presenting it is too skinny or too religious or too conservative or wears funny clothes or has previously presented theories with which you disagree. "Don't believe his theory because he's crazy" is an example of an *ad hominem* fallacy because there is nothing in that statement that says what is wrong with the theory.

The *genetic* fallacy involves refuting a claim because of how it originated. "That argument against evolution is wrong because it came from creationists" and "That claim about the physics of electrons is useless because it was presented by a chemist" are both fallacious because the actual claims are not being scrutinized. In Gray's words, "[w]hether a person's claim is true or false depends solely on whether strong evidence can be marshaled in support of it. And this question of evidence must be sharply distinguished from any other factors that explain why a person makes the claim."[24]

The *appeal to ignorance* fallacy involves either accepting a claim because there is no proof that it is false or rejecting it because it has not been proven true. Absolute proof in science is rare, and most claims should be rejected or accepted tentatively because of our lack of understanding of things. "You cannot prove that global warming will occur in the next century; therefore, it will not occur" and "You cannot prove that aliens from outer space have not visited Earth; therefore, they have been here" are fallacious appeals to ignorance.

When an argument is expected to be accepted because of an *appeal to authority*, a fallacy has been committed. In everyday life, it is common to see authorities in one area being used to support an idea in another area, as in "This top athlete says eat this breakfast cereal" and "I'm a famous entertainer so vote for this politician because I agree with her." In science, even if a leading thinker in a field supports an idea in that field, it is still the idea which has to be evaluated. The opinions of scientific authorities on matters related to their field of expertise can be used to support an idea, but it is wrong to accept an idea only because an authority supports it.

*Two wrongs make a right* is a fallacy in which an act is unacceptably justified by noting that someone else did the same thing. "I eliminated those data points because they did not support my theory, but that is OK because I saw my professor do it once" is an unacceptable argument because both were wrong.

The *bandwagon* fallacy is committed when an act is justified because many people do it or an idea is true because many others believe it. "I know that statistical test is improper for this study, but everyone else uses it, so I will too" and "Aliens from outer space must have visited Earth because millions of people believe in them" are examples of the bandwagon fallacy.

Along the same line is the fallacy of *past practice*, where tradition is used to justify an act or the truth of an idea. "Einstein's theory of relativity can't be right—look how successful Newton's laws have been" or "Astrology has been popular for thousands of years, so it must be true" both avoid the essential issues of the correctness of relativity and astrology.

*Equivocation* is a fallacy in which two meanings of a word or phrase are used without a distinction being made. "Organic foods are healthy, so any food with organic compounds in them must be healthy" is fallacious because the first use of the word *organic* means grown without pesticides, while the second use means things containing carbon-based molecules.

Gray gives a humorous example of equivocation involving the word *average*: "The average family has 2.5 children, and John's family is very average. So, John's family must have 2.5 children."[25]

The fallacies of *division* and *composition* involve the assumptions that what is true of the whole is true of the parts (division), and what is true of parts is true for the whole (composition). "South America has extensive tropical rainforests; therefore, Chile has extensive tropical rainforests" is an example of division. Schick and Vaughn provide a useful example of composition: "Subatomic particles are lifeless. Therefore anything made out of them is lifeless."[26]

Two fallacies are used to disarm an opponent at the beginning of a discussion. *Poisoning the well*, for example, is done in an attempt to make an opponent unable to argue his or her side. "You scientists have closed minds when it comes to the possibility of extra sensory perception, so whatever you say on the issue is meaningless" is an example of poisoning the well. A *loaded question* is worded in such a way as to make it impossible to answer without incriminating oneself. The classic example of a loaded question is "Have you stopped beating your wife?" One scientists might hear is "Are you still wasting taxpayers' money studying useless stuff?" Both yes and no answers suggest that the research has been wasteful and useless.

This is not a complete list of commonly used fallacies, but it does show many of the ways people accept claims for the wrong reasons. When deciding if a claim should be accepted or not, it is important to consider if any of the argument given to support it is fallacious. Of course, this is true for all claims, not just scientific ones.

Good science is the product of clear thinking. The next two chapters deal with pseudoscientific topics—issues presented as scientific that do not meet the standards of science. Pseudoscience exists largely because of a lack of critical thinking and logical reasoning.

### PROBLEMS

1. Write a short essay outlining three misuses of critical thinking

strategies: one by you, one by an acquaintance, and one by a public figure. Your examples need not be related to science.

2. Do the same as in question 1 but with misuses of the fallacies.

3. Why is critical thinking so important for the progress of science?

4. Take one of the scientific norms discussed in Chapter 4 and relate it to a critical thinking strategy or fallacy.

5. Take one example of a form of misconduct discussed in Chapter 5, and relate it to the misuse of a critical thinking strategy or fallacy.

## NOTES

1. Gross, Paul R., and Levitt, Norman, 1994, *Higher Superstition: The Academic Left and Its Quarrels with Science*, Baltimore: The Johns Hopkins University Press, p. 1.

2. Moore, Brooke Noel, and Parker, Richard, 1995, *Critical Thinking*, 4th ed., Mountain View, California: Mayfield Publishing Company, p. 4.

3. Kurfiss, Joanna Gainen, 1988, *Critical Thinking: Theory, Research, Practice, and Possibilities*, ASHE-ERIC Higher Education Report no. 2, Washington, D.C.: Association for the Study of Higher Education, p. 20; italics in original.

4. Stegner, Wallace, 1962, *Beyond the Hundredth Meridian: John Wesley Powell and the Second Opening of the American West*, Sentry Edition, Boston: Houghton Mifflin Co., p. 214–15.

5. Beveridge, W. I. B., 1980, *Seeds of Discovery: A Sequel to The Art of Scientific Investigation*, New York: W. W. Norton & Company, p. 5

6. Paul, Richard W., 1992, "Strategies: Thirty-Five Dimensions of Critical Thinking," in *Critical Thinking: What Every Person Needs to Survive in a Rapidly Changing World*, ed. A. J. A. Binker, Rohnert Park, California: Foundation for Critical Thinking, p. 391–445.

7. Daly, William T., 1995, *Beyond Critical Thinking: Teaching the Thinking Skills Necessary to Academic and Professional Success*, National Resource Center for the Freshman Year Experience and Students in Transition, University of South Carolina, Monograph Series Number 17, p. 5.

8. Twain, Mark, "The Bee," in *The Complete Essays of Mark Twain*, ed. Charles Neider, Garden City, New York: Doubleday and Company, Inc., 1963, p. 530–32. Quote is on p. 531–32; and yes, Twain is being facetious.

9. La Farge, Oliver, 1942, "Scientists Are Lonely Men," *Harper's Magazine* (November): 652.

10. Paul, 1992, p. 407.

11. This quote is included in Dukas, Helen, and Hoffmann, Banesh, eds., 1979, *Albert Einstein: The Human Side (New Glimpses from His Archives)*, Princeton: Princeton University Press p. 48.

12. Paul, 1992, p. 409.

13. Huxley, Thomas Henry, 1860, "Letter to Charles Kinsley," in Huxley, Leonard, 1901, *Life and Letters of Thomas Henry Huxley,* vol. 1, New York: D. Appleton and Company, p. 235.

14. Alcock, James E., 1991, "On the Importance of Methodological Skepticism," *New Ideas in Psychology* 9, no. 2: 151–55; quote is on p. 154.

15. Gray, William D., 1991, *Thinking Critically about New Age Ideas*, Belmont, California: Wadsworth Publishing Co., p. 9–10.

16. Paul, 1992, p. 425.

17. Ibid., p. 426.

18. Ibid., p. 432.

19. Russell, Bertrand, 1962, *Essays in Skepticism*, New York: The Wisdom Library, p. 72–73.

20. Paul, 1992, p. 438–39.

21. Ibid., p. 439.

22. Ibid., p. 444.

23. Gray, 1991, p. 50–77; Schick, Theodore, Jr., and Vaughn, Lewis, 1995, *How to Think about Weird Things: Critical Thinking for a New Age*, Mountain View, California: Mayfield Publishing Co., p. 285–91.

24. Gray, 1991, p. 55.

25. Ibid., p. 65.

26. Schick and Vaughn, 1995, p. 287.

CHAPTER 7

# PSEUDOSCIENCE

This is the excellent foppery of the world, that when we are sick in fortune,
often the surfeits of our own behavior, we make guilty of our disasters the sun,
moon, and stars; as if we were villains on necessity; fools by heavenly compul-
sion; knaves, thieves, and treachers by spherical predominance; drunkards,
liars, and adulterers by an enforced obedience of planetary influence; and all
that we are evil in, by a divine thrusting on. An admirable evasion of whore-
master man, to lay his goatish disposition on the charge of a star.

*William Shakespeare, about 1605*[1]

In this quote from *King Lear*, the character Edmund is railing against
people who avoid responsibility for their actions, laying the blame instead on astrological influences. Astrology is one of many subjects labeled
as pseudoscience. *Pseudoscience* is difficult to define precisely, but in
essence it involves things presented as scientific which do not meet the
standards of science. While science is based on drawing conclusions derived from valid empirical evidence, pseudoscientists use other methods
to validate their ideas. Many practitioners of pseudoscience are sincere in
their beliefs, though in some cases it is promoted by charlatans who do
not believe in the phenomena themselves, but use the gullibility of the
public on pseudoscientific matters to gain fame or, more commonly, fortune. Some pseudosciences are well-ingrained aspects of popular and folk
culture, such as astrology and dowsing, while others are relatively new
ideas promoted by one or a few individuals, such as Immanuel
Velikovsky's suggestion that Venus had near collisions with Earth during
historical times. Pseudoscience sometimes has strong religious overtones,
as in the case of creation "science," and sometimes it has governmental

backing, as with the agricultural "innovations" of Lysenkoism. This chapter outlines some examples of pseudoscience before proceeding with a general discussion of the topic in Chapter 8. It is important to note that the following examples are too brief to provide thorough introductions to the topics and the arguments for and against them.

## ASTROLOGY

*Astrology* is the ancient and popular belief that the positions of the moon, some of the planets, and certain stars at the time of birth influence a person's personality traits and various aspects of daily life. Astrologers generally promote their practice as scientific and useful in explaining personalities and in determining good and bad times for various activities.[2] An astrological assessment is typically vague, with much room for interpretation, which makes many people believe that it was made specifically for them when in fact it applies to most people. Like other pseudosciences, however, when astrology is subjected to careful experiments, it fails. General personality traits are often given for each sun sign, which is the astrological period during which a person is born (e.g., Scorpio, Leo, and Capricorn). John McGervey determined the sun signs for thousands of scientists and politicians, two professions for which certain personality traits are emphasized, yet he found no statistical tendency for members of either profession to be born under some signs more than others.[3]

Shawn Carlson conducted a controlled experiment on the effectiveness of astrology in which prominent astrologers helped to design the experiment.[4] Astrologers were given the results of a psychological test used to measure personality traits for three subjects, and the "natal chart" (which shows the relative locations of the relevant planets, stars, and the moon at the time and place of birth) of one of the three. If astrology works, then the astrologers should have been able to correctly match the natal chart and the personality test results of that individual with reasonable accuracy. The experimental results were that the astrologers made correct matches one-third of the time, which is the same as random guessing. Carlson concluded that ". . . [d]espite the fact that we worked with some of the best astrologers in the country, recommended for their expertise in astrology and for their ability to use the [psychological test used], despite the fact that every reasonable suggestion made by the advising astrologers was worked into the experiment, despite the fact that the astrologers approved the design . . . , astrology failed to perform at a level better than chance. . . . The experiment clearly refutes the astrological hypothesis."[5]

## DOWSING

*Dowsing* is the practice of finding underground water with the use of a tool, often a forked stick, though sometimes a bent wire or a pendulum is used. Dowsers typically walk around an area holding the stick tightly until it points down, which is interpreted as a good place to dig a well. Dowsing, also called water witching and water divining, appears to have originated in mining districts of eastern Europe during the Middle Ages as a means of finding minerals; locating water well sites is an outgrowth of the mining uses. Many thousands of productive wells have been located with the services of dowsers, but when dowsing is subjected to careful scientific scrutiny, it fails experimental tests. One reason for dowsing's many "successes" in finding well sites is that in many areas, groundwater is distributed relatively evenly under the ground surface. Evon Vogt and Ray Hyman give an example of a county in Alabama where dowsers have a 100 percent success rate. However, the entire county has groundwater at about the same depth, and all wells dug without the advice of a dowser are also successful.[6] Where dowsing is used in areas with more complex underground situations, where "dry holes" are common, it generally is found to be no more successful than drilling without a dowser's advice. L. Keith Ward, for example, found that of 1823 dowsed and 1758 nondowsed wells drilled in New South Wales, Australia, 14.7 percent of the dowsed wells were dry, compared to 7.4 percent of the nondowsed wells; having twice the failure rate when dowsing is used does not support dowsers' claims that they help find useful well sites.[7]

When dowsers are subjected to scientifically controlled experiments, they are usually no more effective at finding water than someone making random guesses. James Randi tested four Italian dowsers, who, before the test, all felt that they could find buried water pipes with nearly perfect accuracy, and all agreed to the experimental design. Three pipes were buried in an irregular pattern in a ten meter by ten meter square, and each dowser was given the opportunity to mark the locations of the one pipe with water flowing in it during that test. All four failed completely to locate the pipes.[8]

What makes the dowsing rods move without a conscious effort on the part of the dowser? The movement is likely due to the way in which the rod is held, which allows very slight muscle movements to make it move rapidly and seemingly uncontrollably. When such muscle movements are subconsciously triggered (called ideomoter actions), the movement of the rod may appear to be caused by an outside force.[9]

## WORLDS IN COLLISION

In 1950, Immanuel Velikovsky published *Worlds in Collision*, a book on purported astronomical events and their record in ancient historical accounts.[10] In it, Velikovsky suggests that the planet Venus began as a comet made of material ejected from Jupiter and that it nearly collided with Earth around 1450 B.C. (and other times as well) and that at a later time it became a planet in its present orbit. He interpreted evidence from numerous cultures around the world as records of such collisions and near collisions, especially the Old Testament book of Exodus. The comet, according to Velikovsky, explains the plagues brought upon the Pharaoh of Egypt and the parting of the Red Sea, which allowed the Israelites to escape. Also in Exodus is the story of how, for forty years, the Israelites ate bread, called *manna*, which fell from heaven.[11] Velikovsky suggests that the manna actually came from the comet.[12] In *Worlds in Collision*, Velikovsky presents evidence of astronomical causes for many historical and mythological events, including the biblical story in Joshua of how ". . . the sun stood still, and the moon stayed, until the people had avenged themselves upon their enemies."[13] This event, according to Velikovsky, was caused by a collision between Earth and Venus, which temporarily halted Earth's rotation.[14]

In *Worlds in Collision* and several later books, Velikovsky presented conclusions directly opposed to the mainstream theories of astronomy, geology, archeology, and history, and he gained a devoted following for his ideas among the nonscientific public. There is nothing wrong with arguing against currently held theories in any or all disciplines per se, but many of Velikovsky's ideas were found to stand on very weak scientific and historical foundations. For example, Carl Sagan, J. Derral Mulholland, and David Morrison presented papers in a symposium on Velikovsky's ideas.[15] They each showed, using extremely reliable physical principles, that much of the astronomical basis of Velikovsky's ideas are impossible. Sagan also shows how the myths and legends used by Velikovsky could have arisen without planetary cataclysms. In addition, Peter Huber uses Babylonian historical documents to show that Venus was seen as a planet in its present orbit centuries before 1450 B.C., when Velikovsky claims there was a near collision of the comet Venus with Earth. Velikovsky never felt a need to revise his ideas in light of what most would consider reasonable criticism. According to Henry Bauer, "[s]o convinced was Velikovsky of the correctness of his views that he would not admit, or perhaps did not recognize, some of the points raised as criticisms."[16] (Problems related to the attack on Velikovsky by scientists will be discussed in the next chapter.)

### LYSENKOISM

Governments can promote pseudoscience, as in the case of *Lysenkoism*, which strongly affected the agricultural and biological sciences in the Soviet Union from the 1930s to the 1960s, with disastrous consequences. Trofim Lysenko was an agronomist who used poor scientific reasoning and practice to promote his agricultural innovations, which generally failed. However, he had the strong support of the Communist Party in the years of Stalin and Khrushchev, so he was able to have his ideas put into national policy. He also was able to silence his critics. Those who vocally disagreed with him were fired from their jobs and many were jailed; some died in prison. Lysenko rejected much of modern biology, especially genetics, and had the power to force his ideas on the scientific community in the Soviet Union.[17]

Some of Lysenko's more outlandish theories include cluster plantings and the transformation of species. He decided that young plants do not compete with others of the same species and that if seeds are planted in clusters, most will sacrifice themselves for the good of the species and let one or a few survive. Apparently, Lysenko believed, in true communist fashion, that plants seek a higher good than self-preservation. Without real experimental evidence to support his theory, he declared that it was experimentally confirmed. The Stalin Plan for the Transformation of Nature involved planting vast belts of forest in the grasslands of the southern Soviet Union to alter the climate and make the region better for agriculture. Lysenko's cluster planting method was used, and the disastrous results may have cost the Soviet government a billion rubles in the 1950s (this amount can be roughly converted to well over one billion U.S. dollars).[20]

---

After gaining power, Lysenko turned to open and merciless struggle against all who did not accept his innovations or who tried to follow an independent line in science. Although the Communist Party dictated thought in the social sciences beginning in the 1920s, the natural sciences enjoyed one more decade of relative freedom. Biology then became the first to experience the full weight of ideological control. Today [it has been acknowledged] that the person who, in the name of the Party, conducted a wholesale slaughter in biology and agronomy was Trofim Lysenko. Research and teaching about hormones, neurophysiology, cell theory, genetics, and other biological disciplines were senselessly and fiercely crushed.

Valery Soyfer, 1994[18]

---

Lysenko also had an unusual view of evolutionary biology. Unlike Darwin's suggestion that species arise through a slow process of change, Lysenko claimed that species can be transformed into other species quickly. He believed that if a plant lives in an environment where it is poorly suited to survive, within that plant, seeds or buds of a better adapted species will develop, and the new species will begin to grow. Evidence to support this theory came from a contest in a Lysenko-controlled journal to see who could report the most transformations. Numerous reports followed of wheat transforming into rye, barley to oats, pines to firs, and many others. According to Zhores Medvedev, "[a]ll of these communications were utterly without proof, methodologically illiterate, and thoroughly unreliable. The authors had one leading thought—to please Lysenko, to support by every means the theory advanced by him, to keep the Lysenko trend from being discredited."[21] The fact that crop seeds delivered to farms often were contaminated with seeds from other plants is a more reasonable explanation for the appearance of other species in the fields.[22]

---

Distortion of facts, demagoguery, intimidation, dismissal, reliance on authorities, eyewash, misinformation, self-advertising, repression, obscurantism, slander, fabricated accusation, insulting name calling, and physical elimination of opponents—all were part of the rich arsenal of effective means by which, for nearly thirty years, the "progressive" nature of scientific concepts was confirmed.

Zhores Medvedev, 1969[19]

---

Lysenko had many theories on how to increase agricultural production, and instead of testing the theories experimentally, he had them used on thousands of collective and state farms in the Soviet Union. He would then have the leaders of the farms fill out a questionnaire on how the methods were applied and what the yields were. Many of the people filling out the questionnaires assumed that it was in their best interest to report improved yields, regardless of the actual harvest. This allowed Lysenko to announce tremendous improvements in Soviet farming due to his ideas, even though the results he reported had little or no resemblance to actual agricultural output. For example, the questionnaire method showed that a new method of applying fertilizer was highly successful, yet there was no increase in actual yield, just the yield as reported by those filling out the questionnaires.[23]

Lysenko rejected most of the methods of both research and discourse in science and so can be labeled a pseudoscientist. Lysenkoism is a terrifying example of the consequences of totalitarian control on science and, consequently, the need for academic freedom for all scientists. A highly inefficient agricultural system in the Soviet Union (and in Russia since the breakup of the U.S.S.R.), with serious nutritional and economic consequences for the people, are in part due to the implementation of Lysenko's pseudoscientifically justified farm policies.

Lysenkoism is a good example of what happens when distinctions of good and bad science are made on political grounds instead of scientific ones. Albert Einstein, out of concern for the lack of freedom of thought in the Soviet Union and of the folly of declaring scientific truth to fit a political worldview, commented that "In the realm of the seekers after truth there is no human authority. Whoever attempts to play the magistrate there founders on the laughter of the Gods."[24] He made a similar point with the following poem:

---

**Wisdom of Dialectical Materialism**

By sweat and toil unparalleled
At last a grain of truth to see?
Oh Fool! to work yourself to death.
*Our* party makes truth by decree.
Does some brave spirit dare to doubt?
A bashed-in skull's his quick reward.
Thus teach we him, as ne'er before,
To live with us in sweet accord.

                              Albert Einstein, 1953[25]

---

## CREATION "SCIENCE"

*Creationism* involves the belief that the Old Testament book of Genesis provides an accurate description of Earth history, including plant, animal, and human history. Genesis is interpreted as proof that Earth is six to ten thousand years old; that fossils are remnants of the Great Flood, which occurred during Noah's time; and that the theory of evolution is wrong because all plant and animal species were created in their present forms. Creationism is associated most closely with certain fundamentalist Christian groups, though most Christian denominations do not promote a literal interpretation of Genesis and readily accept scientific explanations for such phenomena, as shown in the accompanying quotes from the Presbyterian Church and Pope John Paul II. James Skehan, both an ordained minister and a geologist, points out that most biblical scholarship suggests that Genesis was written after much of the rest of the

Old Testament and the first eleven chapters of Genesis (those of most interest to creation "science") were most likely an adaptation of the Babylonian creation myth, modified to serve as an introduction to the story of the Israelite's covenant with God, as told in the Old Testament.[26]

---

[A literal interpretation of the Bible] is in conflict with the perspective on Biblical interpretation characteristically maintained by Biblical scholars and theological schools in the mainstream of Protestantism, Roman Catholicism, and Judaism. Such scholars and, we believe, most Presbyterians find that the scientific theory of evolution does not conflict with their interpretation of the origins of life found in the Biblical literature.

General Assembly of the United
Presbyterian Church, U.S.A., 1982[27]

---

As religious doctrine, creationism is not a threat to science, but many followers of creationism want it taught to students not as religion, but as part of the scientific curriculum.[29] *Creation "science"* is an attempt to show that Genesis is literally correct and that scientific theories which contradict such an interpretation are wrong. Most of the effort of creation scientists has been to discredit the theory of evolution and to show that the Earth is about ten thousand years old at most.[30] Whereas science is based on drawing conclusions from empirical evidence, creation scientists start with a belief that their theory is correct and then seek evidence that they can interpret as corroborating their preconceived ideas and as refuting competing theories. For example, radioactive dating techniques are based on very well-established physical principles, and they have been used to show that Earth is about 4.5 billion years old. Creationist Henry Morris presents several problems with radioactive dating which, he feels, makes the methods useless for scientific work,[31] though geologists Stephen Brush and Brent Dalrymple each show that Morris's points are minor and not problematic, misinterpretations of published research, or simply false.[32]

Considerable effort has been made by creation scientists to show that evolution has not occurred and that life could not have arisen on Earth from natural processes. The scientific evidence that evolution has occurred is overwhelming, though certain details of the process are subject to debate. Creation scientists, such as Henry Morris and Duane Gish, use the fossil record to argue that the theory of evolution is false, because fossils showing the transition from one species to another are not found.[33] Arthur Strahler, however, provides several examples of detailed fossil records of species evolving into new species.[34] Likewise, by showing that there is no comprehensive and widely supported theory for the origin of life on Earth, Gish argues that it must have been divinely accomplished, as described in Genesis.[35] Just because scientists have been unable to conclusively determine the natural process by which life first began does not mean that it did not happen under natural conditions. Of course, it should be noted that even if creation scientists were able to show that the theory of evolution is sufficiently flawed to be abandoned (and they certainly have not done so!), that does not lead to the conclusion that the Genesis version of creation is correct and should be taught in schools. The theory of evolution and the theory that Earth is billions of years old are very well-supported by scientific evidence; the creation science theory is very poorly supported by empirical evidence. The idea of creationism has an important place in religious discussions, but there is no valid reason to consider it an important scientific theory or one which should be taught in school science classes either as a theory of equal stature as, or in place of, evolution.[36]

---

Sacred scripture wishes simply to declare that the world was created by God, and in order to teach this truth it expresses itself in the terms of the cosmology in use at the time of the writer. . . . Any other teaching about the origin and makeup of the universe is alien to the intentions of the Bible, which does not wish to teach how heaven was made but how one goes to heaven.

Pope John Paul II, 1983[28]

---

## UFOS

*Unidentified flying objects* (UFOs) are commonly believed to be of extraterrestrial origin, when in fact they are, as the phrase implies, merely unidentified. Under careful examination, most UFO sitings have a reasonable explanation, such as weather balloons, the planet Venus (which can appear unusually bright at times), and distant aircraft. The actual cause for sightings cannot be determined every time, but that is due to a lack of information and does not lead to the conclusion that space aliens visited the planet.

People who report UFOs sometimes include detailed information on the spaceship "seen." Psychological studies have shown, however, that human memory of an event is not simply a reflection of the reality of what happened, but a combination of the actual event, how the brain processes the event, and later modifications of the memory. This applies to all people, not just to those with psychological disorders. Many people have been frightened by a threatening person lurking in a shadow on a dark night only to find that the "person" really was a shrub.[37] Sometimes, however, people hurry away and never find out that the attacker did not exist. Often in such situations, that person will keep a memory of a reality that never existed, possibly including details of the attacker's sinister face. Similar misinterpretations of reality are common in UFO sitings.

---

The great failure of the pro-UFO movement has been the unwillingness to accept the fact that human perception and memory are not only unreliable under a variety of conditions (and these conditions are exactly those under which most UFOs are reported) but that perception and memory are also *constructive.* That is, perception is a function not only of the actual sensory stimulus that is picked up by the eye or the ear but also a function of what we know and believe about the world, even if that knowledge and belief are wrong. The constructive nature of perception is greatest when the actual sensory input is weak, unclear, or ambiguous— just the type of sensory input present in most UFO sitings. Memory, too, is constructive. Experiments . . . show clearly that what we remember about an incident can actually be changed after the fact. When this happens, the witness truthfully testifies to remembering something that never happened.

Terrence Hines, 1988[38]

---

Vision and memory may at times be unreliable in UFO sitings (and many other situations as well), which may explain many of the reported cases of alien space craft. Some (probably a minority) who report UFOs, however, have psychological problems with delusions, and others report them as hoaxes or as attempts to get attention. Most photographs of UFOs have earthbound explanations or have been shown to be deliberate fakes. Extraterrestrial beings *may* exist, *may* have visited Earth, and some of the reports of UFOs *may* be accurate, but as yet there is no conclusive evidence of such visits.[39]

## PARANORMAL PHENOMENA

*Paranormal phenomena* include a variety of topics for which known physical concepts do not appear to be involved. Such psychic activities as telepathy (mind reading, etc.), precognition ("seeing the future"), faith healing, psychic surgery, psychokinesis (exerting control over inanimate objects by thought alone, e.g., "spoon bending"), and contacting the spirits of the dead are examples of paranormal phenomena, as are ghosts and pyramid power (the notion that "energy" from pyramid-shaped objects can do such things as preserve corpses and even sharpen dull razor blades). Dowsing, discussed earlier, also is a paranormal activity.[40] Paranormal phenomena are generally considered *extrasensory*, in that energy or information is transferred between individuals or between an individual and the environment without the use of the five senses of sight, smell, touch, hearing, or taste. *Psychics* are people who claim to have such powers. Some psychics are sincere in their beliefs, and others are frauds using deception to fool gullible people.[41]

*Psychic detectives* often claim credit for using their powers to help police solve crimes. To check such claims, Jane Ayers Sweat and Mark Durm sent a questionnaire to the police departments of the fifty largest cities in the United States asking about the use of psychics. Most had never used psychics, and the respondents overwhelmingly felt that information provided by psychics was no more useful than other information. The response from Austin, Texas, included the comment that "I have yet to see any information received from psychics of any value, based on 20 years experience. The information is usually distorted, of no investigative value, and inaccurate. They hamper an investigation and often cause distractions from the main investigation."[42] From Portland, Oregon: "Psychic information has been volunteered many times, but has *never* been beneficial to a case."[43] Controlled tests of the abilities of psychic detectives to provide accurate statements about crimes show that they are no more correct than "nonpsychics."[44] Many claims of success by psychic detectives are simply untrue.[45]

Predicting the future by psychic means would require a transfer of information from the future to the predictor. Given our present understanding of physics, such a delivery is impossible.[46] Nostradamus is credited as the greatest psychic predictor of future events. In the sixteenth century, he wrote a series of poems which have been interpreted as foretelling of subsequent events in world history. Unfortunately, the "predictions" typically have been identified after the fact, not before the event occurred. Nostradamus's poems are vague and allow much room for dubious interpretation and questionable translation by people eager to find certain events foretold in them.[47] One of the poems is translated as

> Animals ferocious with hunger will swim the rivers,
> The greater part of the armed camp will be against Hister;
> The great man will be carried in an iron cage,
> When the German child watches the Rhine[48]

This poem was interpreted by Stewart Robb as a prediction of the end of Adolph Hitler's power, because of the war theme and because in Roman times the Lower Danube was known as "Hister," which has a similar spelling to "Hitler."[49] A more reasonable explanation, provided by James Randi, is that Nostradamus was referring to Turkish forces, which had advanced into Europe to the Lower Danube at the time the poem was written.[50] To his credit, Robb (who begins his book with "Prophesy is a scientific fact . . . "[51]) published his interpretation of this poem before Hitler's defeat, which did occur, though the "iron cage" is interpreted as prison, while Hitler actually committed suicide before capture.

Most paranormal phenomena have normal explanations. For example, many people have had the experience of thinking of a friend or relative and then unexpectedly getting a phone call from that person. Such an event is often interpreted as evidence of extrasensory powers. However, a more reasonable explanation is that it is a coincidence. What we fail to do is compare the unusual event with all of the more common events of thinking of that person without a subsequent phone call or getting a call without thinking about him or her beforehand.[52] The *principle of parsimony*, also known as *Occam's razor*, is described by Arthur Strahler as the idea that we should not ". . . bring in unneeded explanations or causes if those we already have on hand are adequate to do the job."[53] It is, therefore, unnecessary to use paranormal explanations for phenomena that casily can be explained using established scientific knowledge.

Fortune tellers, mind readers, and other assorted psychics have a variety of tricks used to make people believe that they have extrasensory powers. These tricks range from giving vague statements about a person and, using clues from facial expressions, then making the statements more specific, to using magic tricks to make it look like they can bend spoons using mental powers. Highly general statements about the location of a missing body or someone's personal problems are often interpreted as specific psychic insight after the information is revealed by nonpsychic means.[54]

Belief in paranormal phenomena is strong among the public. The paranormal suggests new dimensions of reality that we have failed to appreciate and untapped powers that we all have, which might explain why the will to believe exists in many. The idea of extraterrestrial visitors adds a similar mysterious and intriguing quality to our existence. As such, these things have long been staples of fiction, from ghost stories to heroes with superhuman powers to visitors from space. Because they make such popular novels, movies, and television shows, paranormal phenomena and UFOs are frequently used in "nonfiction" as well. Unfortunately, their storytelling appeal typically eclipses the quest for truth in nonfiction. It is common for the phenomena to be presented in a way such that the paranormal is accepted as fact. Even reputable news organizations frequently suspend their journalistic skepticism in order to make stories more "interesting." James Randi reports on a "prediction" of an airline crash in 1977 in which a description of the crash was sealed in an envelope and locked in a safe before the accident. The person making the prediction clearly stated that he had no supernatural powers but was, instead, a professional magician. (It is the magician's job to make the impossible appear to happen.) Only one newspaper out of seventeen carrying the story included that information; the rest reported it as a psychic prediction.[55] This fits with journalist Phillip Meyer's comment that "[t]he old maxim 'Don't let the facts get in the way of a good story' is especially tempting when reports of the paranormal are concerned. The human desire to believe is so strong and the stories so intrinsically interesting that the temptation to suspend critical judgment is almost irresistible."[56] It is not surprising that belief in the paranormal and UFOs is so popular, considering the commonly held will to believe; the barrage of such things in fictional and nonfictional television, movies, and books; and the frequent appearance of uncritical reports in the news media.

*Parapsychology* is a branch of science involved with the study of paranormal activity, especially extrasensory perception (ESP) and psychokinesis.[57] Parapsychologists have performed many experiments to test for the existence of paranormal powers (or *psi* phenomena, as they are sometimes called). Many of the tests have shown no evidence of psi phenomena, while many have concluded that the existence of psi was shown. Of those with positive results (showing paranormal activity), many can be criticized for flawed experimental design, which could allow for incorrect results, and some can be shown to involve fraud in the experimenting. Some of the positive results, however, cannot easily be explained away, which can be interpreted as *possible* evidence for the existence of psi. What is missing, in the minds of most skeptics of psi research, is repeated experiments from which the same positive results are obtained. To provide reliable scientific evidence of psi phenomena, the experiments should be rigorously designed and executed. They also should be repeated several times by other researchers, with consistent results. After decades of research and thousands of experiments, such evidence has not been provided by parapsychology researchers.[58]

## HEALTH-RELATED PSEUDOSCIENCE

*Medical quackery* is common and often dangerous. Many pseudoscientific medical procedures are useless or harmful, and many fad diets are nutritionally unsound. As an example of the latter, in *Fit for Life*, authors

Harvey and Marilyn Diamond make the suggestion that fats, carbohydrates, and proteins be eaten at separate meals, because certain combinations of food types putrefy in the body and cause unhealthy toxins. They also recommend eating foods with a high water content to flush these toxins out of the body.[59] Nutrition research shows that there is no scientific basis for such assertions and that following the recommended diet for long periods of time would lead to nutritional deficiencies. William Jarvis commented on *Fit for Life:* "It is appalling that such a book can become a best seller in the latter half of the twentieth century. Its only socially redeeming feature is that its popularity may alert American educators to their failure to impart the most fundamental knowledge about health and nutrition to the students entrusted to their care."[60]

An example of medical pseudoscience is *homeopathy,* which is based on the premise that if a substance causes a disease, then extremely small doses of that substance can cure the disease. Homeopathic substances are diluted in water to such an extent that often it is unlikely that a single molecule of it is left in the water drunk by the patient, though homeopathic practitioners claim that an "essence" of the substance remains. The success of homeopathy likely is due to the fact that most diseases, especially minor ones, can be cured by the body without medical treatment and to the "placebo effect," in which a substance has no pharmacological value but is given to a patient who believes it will help. If the problem is psychosomatic (medical problems which are not completely physiological but partly psychological in nature), then placebos may help. However, homeopaths claim that their treatments are not placebos, but have direct physiological effects. While homeopathy contradicts established knowledge in chemistry, physics, and pharmacology and rigorous scientific studies suggest that it does not work as claimed, it has been popular for nearly 200 years.[61] Sales of homeopathic products in 1992 were $160 million by one estimate and $200 million in another.[62]

Medical quacks are common and typically gain their patients' confidence with promises of cures for everything from cancer to AIDS to baldness and impotence, usually for a price, of course. Special diets or drugs or vitamins may be prescribed without any evidence of their effectiveness. Laetrile, for example, was actively promoted as an effective cancer treatment, mainly in the 1960s and 1970s. Without any scientific support for claims that it is effective at controlling cancer, thousands of people in the United States spent millions of dollars on the drug.[63] As AIDS spread in the 1980s, many quack cancer treatments were promoted as AIDS treatments without any evidence that they helped either malady.[64]

A successful selling point for medical quackery is the testimonial of a "cured" patient. As Martin Gardner commented, there are two reasons for the success of quackery, as mentioned previously with respect to homeopathy. One is that many medical problems, even serious ones, disappear without treatment at least some of the time. Another is that many ills have a psychosomatic component. Therefore, it is relatively easy to find patients who believe that they were cured by procedures that had no real effect on the disease.[65] It is also common to find people "cured" of a disease, who in fact only feel better temporarily. Unfortunately, some people who mistakenly believe they have been cured of a disease stop conventional medical treatment and die.[66]

---

Ultimately, answering fundamental questions about efficacy, safety, appropriate clinical applications, and meaningful outcomes for all medical therapies, including those considered alternative medicine, requires critical and objective assessment using accepted principles of scientific investigation and rigorous standards for evaluation of scientific evidence. For patients, for physicians and other health care professionals, and for alternative medicine practitioners—indeed, for all who share the goal of improving the health of individuals and of the public—there can be no alternative.

Phil B. Fontanarosa and
George B. Lundberg, 1998[69]

---

Many health-related pseudosciences are called *alternative medicines*, but not all alternative medicines are necessarily pseudoscience. Alternative medicine is difficult to define precisely but includes those practices which are not part of conventional Western-style medical practice. Some, like laetrile, are clearly pseudoscientific, while others, like acupuncture and chiropractic, are less easily categorized. The general problem with most alternative medicines is that they are accepted as useful, though they have not been tested sufficiently. Rigorous scientific experiments on the value of these treatments is needed to determine which are useful and which are mere placebos or scams.[67] Of course, scientific testing should be done for all conventional medical treatments, too, which is not always the case.[68]

SOME OTHER PSEUDOSCIENCES

Other examples of pseudoscience include cryptozoology, Atlantis, graphology, the lunar effect, and the Bermuda Triangle. *Cryptozoology* is the study of mythical creatures which have not been identified by science. These include Bigfoot (also called Sasquatch), Yeti (or the Abominable Snowman), and the Loch Ness Monster. Much like UFOs, these creatures may exist, but the evidence is insufficient to warrant acceptance of them as real.[70] The lost continent of *Atlantis* is believed to have sunk into the ocean. Atlantis occasionally is "found" by pseudoscientists, but there is no reliable evidence of such a disappearing continent.[71] *Graphology* is used to gain information about people from their handwriting style. Some personal attributes, such as gender, can be determined from handwriting with reasonable accuracy, but the practice is used in ways for which no evidence exists for its reliability. Graphology is used by thousands of companies in making personnel decisions, such as determining

---

The Legend of the Bermuda Triangle is a manufactured mystery. It began because of careless research and was elaborated upon and perpetuated by writers who either purposely or unknowingly made use of misconceptions, faulty reasoning, and sensationalism. It was repeated so many times that it began to take on the aura of truth.

Larry Kusche, 1995[75]

---

the honesty of a job applicant, even though it cannot be shown to be accurate for such purposes.[72] The *lunar effect* is the idea that the full moon affects human behavior in such ways as increasing rates of crime, human births, suicides, and admissions to mental hospitals. The word *lunacy* comes from this belief. Studies, however, show no objective basis for these claims.[73] The *Bermuda Triangle* is a section of the western Atlantic

Ocean where, according to Charles Berlitz ". . . more than 100 ships and planes have literally vanished into thin air."[74] Ships and planes do disappear there mysteriously, but no more frequently than any other section of ocean with similar traffic and weather patterns.

Many more examples of pseudoscience could be given here. "Flat Earthers"[76] and perpetual motion machines[77] have not even been mentioned, much less the practice of eating tiger penis soup (at $320 per serving) to enhance sexual performance,[78] but this chapter has to end sometime.

Pseudoscience ranges from the silly to the deadly. It has been said that "[t]here is just one basic joke in all humor: human beings trying to impose their will on reality and reality showing them up for fools . . . "[79] From this perspective, followers of pseudoscience are very funny, but considering the misery that sometimes results, the humor can be very dark indeed. The next chapter will discuss pseudoscience in more general terms.

## PROBLEMS

1. Choose a pseudoscience and design an experiment to test it. Be sure to start with a clearly stated hypothesis.

2. List several critical thinking strategies or fallacies that relate to one of the pseudosciences discussed here.

3. How can scientists help journalists reporting on pseudosciences?

## NOTES

1. Shakespeare, William, 1963, *The Tragedy of King Lear*, Signet Classic edition, New York: New American Library, p. 56. The play was written near the beginning of the seventeenth century, and the quote is from scene 1, act 2.

2. Royer, Mary-Paige, 1991, *Astrology: Opposing Viewpoints*, San Diego: Greenhaven Press, p. 12–45.

3. McGervey, John D., 1981, A Statistical Test of Sun-Sign Astrology, in *Paranormal Borderlands of Science*, ed. Kendrick Frazier, Buffalo: Prometheus Books, p. 235–40.

4. Carlson, Shawn, 1985, "A Double-Blind Test of Astrology," *Nature* 318: 419–25.

5. Ibid., p. 425.

6. Vogt, Evon Z., and Hyman, Ray, 1979, *Water Witching, USA*, 2d ed., Chicago: University of Chicago Press, p. 37.

7. Ward, L. Keith, 1951, "Search for Underground Water Supplies (Underground Water in Australia—8)," *Chemical Engineering and Mining Review* 43 (May 10, 1951): 291–97.

8. Randi, James, 1982, *Flim Flam!: Psychics, ESP, Unicorns and Other Delusions*, Buffalo: Prometheus Books, p. 307–24. The experimental design in the dowsing study is more complex than presented here.

9. Vogt and Hyman, 1979, p. 121–52.

10. Velikovsky, Immanuel, 1950, *Worlds in Collision*, New York: Pocket Books.

11. *The Holy Bible in the King James Version*, 1984, Nashville, Tennessee: Thomas Nelson Publishers. See Exodus 7 though 16, p. 39–47.

12. Velikovsky, 1950, p. 69–74.

13. *Holy Bible*, Joshua 10:13, p. 147.

14. Velikovsky, 1950, p. 55–60.

15. Goldsmith, Donald, ed., 1977, *Scientists Confront Velikovsky*, Ithaca, New York: Cornell University Press. Individual chapters cited are Carl Sagan, "An Analysis of *Worlds in Collision*," p. 41–104; J. Derral Mulholland, "Movements of Celestial Bodies—Velikovsky's Fatal Flaw," p. 105–15; Peter J. Huber, "Early Cuneiform Evidence for the Existence of Planet Venus," p. 117–44; and David Morrison, "Planetary Astonomy and Velikovsky's Catastrophism," p. 145–76.

16. Bauer, Henry H., 1984, *Beyond Velikovsky: The History of a Public Controversy*, Urbana: University of Illinois Press, p. 163.

17. Soyfer, Valery N., 1994, *Lysenko and the Tragedy of Soviet Science*, New Brunswick, New Jersey: Rutgers University Press, p. xviii.

18. Ibid., p. 2.

19. Medvedev, Zhores A., 1969, *The Rise and Fall of T. D. Lysenko*, New York: Columbia University Press, p. 166–70.

20. Soyfer, 1994, p. 205-9. See also Medvedev, 1969, p. 191–92.

21. Medvedev, 1969, p. 170.

22. Ibid., p. 170–72. See also Soyfer, 1994, p. 209–11.

23. Medvedev, 1969, p. 153–55.

24. Hoffmann, Banesh, and Dukas, Helen, 1972, *Albert Einstein: Creator and Rebel*, New York: Viking Press, p. 245–46.

25. Hoffmann and Dukas, 1972, p. 246; italics in original.

26. Skehan, James W., 1983, "Theological Basis for a Judeo-Christian Position on Creationism," *Journal of Geological Education* 31, p. 307–14.

27. Resolution and Supporting Information on Evolution and Creationism, adopted June 28, 1982, by the 194th General Assembly of the Presbyterian Church, U.S.A.; quoted in Frye, Roland Mushat, 1983, "Creation-Science Against the Religious Background," in *Is God a Creationist?: The Religious Case Against Creation-Science*, ed. Roland Mushat Frye, New York: Charles Scribner's Sons, p. 7. Frye (p. 4–5) also quotes a statement of similar sentiment from the Episcopal Church of the United States.

28. Pope John Paul II, 1983, "Science and Christianity," in Frye, 1983, p. 153. According to Frye (p. 8), the Pope is intentionally paraphrasing a statement made by Galileo while

on trial before the Inquisition, and so ". . . provides a graceful, yet forceful warning against repeating that tragic seventeenth-century effort to dominate science by a literalistic interpretation of Genesis."

29. Gish, Duane T., 1995, *Teaching Creation Science in Public Schools*, El Cajon, California: Institute for Creation Research, p. 67.

30. Presented here is a highly simplified version of creation science, and the reader should be aware that there are several varieties of it. See, for example, Scott, Eugenie C., 1996, "Creationism, Ideology, and Science," in *The Flight from Science and Reason*, eds. Paul R. Gross, Norman Levett, and Martin W. Lewis, New York: New York Academy of Sciences, p. 505–22.

31. Morris, Henry M., ed., 1974a, *Scientific Creationism*, gen. ed., San Diego: Creation-Life Publishers, p. 131–49.

32. Brush, Stephen G., 1982, "Finding the Age of the Earth by Physics or by Faith?" *Journal of Geological Education* 30: 34–58. See also Dalrymple, G. Brent, 1984, "How Old is the Earth? A Reply to 'Scientific' Creationism," in *Evolutionists Confront Creationists*, eds. Frank Aubrey and William M. Thwaites. Proceedings of the 63rd Annual Meeting of the Pacific Division, American Association for the Advancement of Science, Part 1, Pacific Division, American Association for the Advancement of Science, San Francisco, p. 66–131.

33. Morris, Henry M., 1974b, *The Troubled Waters of Evolution*, San Diego: Creation-Life Publishers, p. 79–109. See also Gish, Duane T., 1993, *Creation Scientists Answer Their Critics*, El Cajon, California: Institute for Creation Research, 111–49.

34. Strahler, Arthur N., 1987, *Science and Earth History—The Evolution/Creation Controversy*, Buffalo: Prometheus Books, p. 398–400.

35. Gish, 1993, p. 151–208.

36. See, for example, Working Group on Teaching Evolution, National Academy of Sciences, 1998, *Teaching About Evolution and the Nature of Science*, Washington, D.C.: National Academy Press.

37. Schick, Theodore, Jr., and Vaughn, Lewis, 1995, *How to Think about Weird Things: Critical Thinking for a New Age*, Mountain View, California: Mayfield Publishing Co., p. 40.

38. Hines, Terrence, 1988, *Pseudoscience and the Paranormal: A Critical Examination of the Evidence*, Buffalo: Prometheus Books; italics in original. For more general discussions of the fallibility of memory, see Loftus, Elizabeth F., 1979, *Eyewitness Testimony*, Cambridge: Harvard University Press, p. 253 and Wiseman, Richard; Smith, Matthew; and Wiseman, Jeff; 1995, "Eyewitness Testimony and the Paranormal," *Skeptical Inquirer* 19, no. 6 (November/December), p. 29–32.

39. For example, see Hines, 1988, p. 165–210; Klass, Phillip J., 1989, *UFO Abductions: A Dangerous Game,* updated ed., Buffalo: Prometheus Books.; Bryan, C. D. B., 1995, *Close Encounters of the Fourth Kind: Alien Abduction, UFOs, and the Conference at M.I.T.*, New York: Alfred A. Knopf.

40. For a more complete outline of paranormal phenomena, see Randi, James, 1995, *An Encyclopedia of Claims, Frauds, and Hoaxes of the Occult and Supernatural*, New York: St. Martin's Press.

41. Frazier, 1981.

42. Sweat, Jane Ayers, and Durm, Mark D., 1993, "Psychics: Do Police Departments Really Use Them?" *Skeptical Inquirer* 17: 148–58; quote is on p. 156. This article is reprinted in Nickell, Joe, ed., 1994, *Psychic Sleuths: ESP and Sensational Cases*, Buffalo: Prometheus Books, p. 212–23.

43. Sweat and Durm, 1993, p. 157–58; italics in original.

44. Wiseman, Richard; West, Donald; and Stemman, Roy; 1996, "Psychic Crime Detectives: A New Test for Measuring Their Successes and Failures," *Skeptical Inquirer* 20: 38–40, 58.

45. See Sweat and Durm, 1993. See also Rowe, Walter F., 1993, "Psychic Detectives: A Critical Examination," *Skeptical Inquirer* 17: 159–65. This article, too, is reprinted in Nickell, Joe, ed., 1994, *Psychic Sleuths: ESP and Sensational Cases*, Buffalo: Prometheus Books, p. 236–44.

46. Rothman, Milton A., 1988, *A Physicists Guide to Skepticism*, Buffalo: Prometheus Books, p. 150. See also Rothman, Milton A., 1991, *The Science Gap: Dispelling the Myths and Understanding the Reality of Science*, Buffalo: Prometheus Books, p. 80–82.

47. Randi, James, 1993, *The Mask of Nostradamus: The Prophecies of the World's Most Famous Seer*, Buffalo: Prometheus Books.

48. Robb, Stewart, 1941, *Nostradamus on Napoleon, Hitler and the Present Crisis*, New York: Charles Scribner's Sons, p. 171.

49. Ibid., p. 168–72.

50. Randi, 1993, p. 212–14.

51. Robb, 1941, p. 3.

52. Alcock, James E., 1995, "The Belief Engine," *The Skeptical Inquirer* 19 (May/June): 14–18. See also Gilovich, Thomas, 1991, *How We Know What Isn't So: The Fallibility of Human Reason in Everyday Life*, New York: The Free Press, p. 174–80.

53. Strahler, Arthur N., 1992, *Understanding Science: An Introduction to Concepts and Issues*, Buffalo: Prometheus Books. "Occam's razor" is named for William of Occam, a fourteenth-century philosopher who is credited with the statement "Entities must not be multiplied beyond necessity." For more detailed discussions of Occam's razor, see Strahler, 1992, p. 50–51 and Rothman, 1988, p. 172–75.

54. See, for example, the collection of papers in "Book I: Skeptical Inquiries into the Paranormal," in Frazier, 1981.

55. Randi, James, 1977, "The Media and Reports of the Paranormal," *The Humanist* 37, no. 4 (July/August): 45–47.

56. Meyer, Phillip, 1986, "Ghostboosters: The Press and the Paranormal," *Columbia Journalism Review* 24: 38–41; quote is on p. 39.

57. For an overview of the field of parapsychology, see Wolman, Benjamin B., ed., 1977, *Handbook of Parapsychology*, New York: Van Nostrand Reinhold Company, p. 967. For a discussion of how parapsychology is treated by other scientists, see McClenon, James, 1984, *Deviant Science: The Case of Parapsychology*, Philadelphia: University of Pennsylvania Press. See also Kurtz, Paul, 1985, "Is Parapsychology a Science?" in *A Skeptic's Handbook of Parapsychology*, ed. Paul Kurtz, Buffalo: Prometheus Books, p. 503–18 and Bunge, Mario, 1991, "A Skeptic's Beliefs and Disbeliefs," *New Ideas in Psychology* 9: 131–49.

58. See Hyman, Ray, 1985, "A Critical Overview of Parapsychology," in *A Skeptic's*

*Handbook of Parapsychology*, ed. Paul Kurtz, Buffalo: Prometheus Books, p. 3–96. See also Hansel, C. E. M., 1989, *The Search for Psychic Power: ESP & Parapsychology Revisited*, Buffalo: Prometheus Books. See also Hines, Terence, 1988, *Pseudoscience and the Paranormal: A Critical Examination of the Evidence*, Buffalo: Prometheus Books, p. 21–108.

59. Diamond, Harvey, and Diamond, Marilyn, 1985, *Fit for Life*, New York: Warner Books.

60. Barrett, Stephen, 1993, "Weight Control: Facts, Fads, and Frauds," in *The Health Robbers: A Close Look at Quackery in America*, eds. Stephen Barrett and William T. Jarvis, Buffalo: Prometheus Books, p. 113–24; Jarvis quote on p. 117.

61. Jarvis, William, ed., 1995, "Homeopathy: A Position Statement by the National Council Against Health Fraud," *Skeptic* 3, no. 1: 50–57. See also Gardner, Martin, 1957, *Fads and Fallacies in the Name of Science*, 2d ed., New York: Dover Publications, p. 187–91 and Gardner, Martin, 1992, *On the Wild Side*, Buffalo: Prometheus Books, p. 30–40. Also see Barrett, Stephen, 1993, "Homeopathy: Is It Medicine?" in *The Health Robbers: A Close Look at Quackery in America*, eds. Stephen Barrett and William T. Jarvis, Buffalo: Prometheus Books, p. 191–202.

62. Barrett, Stephen, and Herbert, Victor, 1994, *The Vitamin Pushers: How the "Health Food" Industry Is Selling America a Bill of Goods*, Buffalo: Prometheus Books, p. 173.

63. Arje, Sidney L., and Smith, Lois V., 1976, "The Cruellest Killers," in *The Health Robbers: How to Protect Your Money and Your Life*, eds. Stephen Barrett and Gilda Knight, Philadelphia: George F. Stickley Company, p. 1–13. See also Young, James Harvey, 1992, *American Health Quackery*, Princeton: Princeton University Press, p. 205–55.

64. Young, 1992, p. 256–85

65. Gardner, 1957, p. 186–87; see also Young, James Harvey, 1993, "Why Quackery Persists," in Barrett and Jarvis, 1993, p. 457–64.

66. Kingslover, Lester, and Barrett, Stephen, 1993, "The Miracle Merchants," in Barrett and Jarvis, 1993, p. 337–45. See also Randi, James, 1989, *The Faith Healers*, new updated ed., Buffalo: Prometheus Books.

67. Fontanarosa, Phil B., and Lundberg, George B., 1998, "Alternative Medicine Meets Science," *Journal of the American Medical Association* 280, no. 18: 1618–19.

68. See, for example, Taubes, Gary, 1998, "The (Political) Science of Salt," *Science* 281: 898–907.

69. Fontanarosa and Lundberg, 1998, p. 1619.

70. Hines, 1988, p. 284–89; see also Gaffron, Norma, 1989, *Bigfoot: Opposing Viewpoints*, San Diego: Greenhaven.

71. Gardner, 1957, p. 164–72; see also Vitaliano, Dorothy, B., 1973, *Legends of the Earth: Their Geologic Origins*, Bloomington: Indiana University Press, p. 218–51. It is possible that the legend of Atlantis began with the volcanic eruption of Santorin in the Aegean Sea around 1450 B.C., which likely led to the decline of the Minoan civilization on Crete. This same eruption could have led to some of the stories that Velikovsky attributes to astronomical phenomena.

72. Hines, 1988, p. 294–97; see also Neter, Efrat, and Ben Shakhar, Gershon, 1989, "The Predictive Validity of Graphological Inferences: A Meta-Analytic Approach," *Personality and Individual Differences* 10: 737–45. Their results on the value of grapholog-

ical analysis of job applicants suggest ". . . that graphologists are not better than non-graphologists in predicting future performance on the basis of handwritten scripts" (p. 743) and that when graphologists are accurate, it is more likely to be due to information gained from the content of what was written than to the handwriting style itself.

73. Abell, George O., and Greenspan, Bennett, 1979, "Human Births and the Phase of the Moon," *New England Journal of Medicine* 300: 96. See also Rotton, James, and Kelly, I. W., 1985, "Much Ado About the Full Moon: A Meta-Analysis of Lunar-Lunacy Research," *Psychological Bulletin* 97: 286–306. One explanation offered by skeptics for the popularity of the lunar effect is the idea that nurses and others who believe in it seek out instances of unusual behavior during full moons because they expect to find it and are less likely to note it at other times. However, Monica Angus directly studied this phenomenon and found no difference between believing and nonbelieving nurses in the propensity to note unusual occurrences during full moons. See Angus, Monica Diane, 1973, *The Rejection of Two Explanations of Belief in A Lunar Influence on Behavior*, M.A. Thesis, Simon Fraser University, Burnaby, British Columbia, Canada, p. 33–40.

74. Berlitz, Charles, 1974, *The Bermuda Triangle*, Garden City, New York: Doubleday and Company, Inc., p. 11.

75. Kusche, Larry, 1995, *The Bermuda Triangle Mystery—Solved*, 2d ed., Buffalo: Prometheus Books, p. 277.

76. Gardner, 1957, p. 16–27.

77. Rothman, 1988, p. 19–35

78. Brower, Kenneth, 1994, "Devouring the Earth," *The Atlantic Monthly* 274 (November): 113–26. See also Tedeschi, Chris, 1995, "Snuffing Tigers for Sex and Profit," *Discover* 16, no. 1: 59. Such practices have contributed to the near extinction of several subspecies of tiger.

79. Young, Charles M., 1978, "Terrors of the Universe and other Cartoons," *Rolling Stone* Issue 274 (21 September 1978): 34–37; quote is on p. 35.

# CHAPTER 8

# SCIENCE, PSEUDOSCIENCE, AND A SKEPTICAL ATTITUDE

Even when the experts all agree, they may well be mistaken. Einstein's view as to the magnitude of deflection of light by gravitation would have been rejected by all experts not many years ago, yet it proved to be right. Nevertheless the opinion of experts, when it is unanimous, must be accepted by non-experts as more likely to be right than the opposite opinion. The scepticism that I advocate amounts only to this: (1) that when the experts agree, the opposite opinion cannot be held to be certain; (2) that when they are not agreed, no opinion can be regarded as certain by a non-expert; and (3) that when they all hold that no sufficient grounds for a positive opinion exist, the ordinary man would do well to suspend his judgment.

These propositions may seem mild, yet, if accepted, they would absolutely revolutionize human life.

*Bertrand Russell, 1928*[1]

The examples of pseudoscience presented in the previous chapter suggest its diverse nature, though there are common themes that apply to most pseudoscientific topics.[2] Science, as outlined in the first five chapters of this book, has developed a highly effective system for learning about real phenomena, including checks on the reliability of the knowledge gained. Pseudoscientists typically avoid some or all of the aspects of scientific inquiry yet present their claims as established scientific facts. An understanding of science and of critical thinking leads to the development of a skeptical attitude, which is the most successful approach to evaluating the claims of both scientists and pseudoscientists.

## SCIENCE AND PSEUDOSCIENCE

Many pseudosciences are historically tied to sciences. Astronomy, for example, grew out of astrology during the Renaissance as modern scientific methods began to be used to study the stars and planets. Medical practice was mostly superstition until scientific research began to determine treatments that truly worked, a goal still not fully attained. (One advantage of homeopathy when it was first developed in the late 1700s is that it did less harm than many conventional medical treatments of the time.) The biblical story of creation was not doubted by most European scientists until the nineteenth century, when geology and biology adopted a more modern scientific approach. While historically science and pseudoscience often blend together, today it is usually easy to differentiate the two. Science uses the methods outlined in this book, and pseudoscience does not.

Pseudoscience often involves phenomena that contradict currently accepted theories in related areas of science. Many important contributions to science have begun as challenges to established thought and that, by itself, is not a mark of pseudoscience. Isaac Asimov groups those who challenge established scientific thought, whom he calls *scientific heretics*, into endoheritics and exoheritics. *Endoheritics* are members of the scientific community and present their challenges to scientific orthodoxy through established means, such as peer-reviewed journal articles and books. Early in the twentieth century, for example, Alfred Wegener presented his ideas on continental drift, which were contrary to much of the theoretical understanding of earth science at the time. Wegener was ridiculed by many until the 1960s, long after his death, when new evidence supported many of his claims. In an earlier time when science and religion were not separated in Europe, Galileo was brought before the Inquisition for his heretical idea that the Earth revolves around the Sun. Keep in mind, though, that for every endoheritic who, often belatedly, is shown to be correct, there are dozens whose ideas are most likely wrong.[3] Supporters of pseudoscientists, in the words of Carl Sagan, ". . . often point to geniuses of the past who were ridiculed. But the fact that some geniuses were laughed at does not imply that all who are laughed at are geniuses."[4] As Stephen Jay Gould phrases it, "But a man does not attain the status of Galileo merely because he is persecuted; he must also be right."[5]

The self-correcting nature of science is such that if valid ideas are presented and effectively supported with evidence, they eventually should be accepted by the scientific community. In the long run, it is better to reject a few correct ideas rather than accept many incorrect ones. This is an imperfect system and can be cruel to good scientists (though it should not be), but it also is the best way to improve scientific knowledge. Well-established theories should not be discarded easily.[6]

---

There is no way, then, of dealing with the endoheresies other than by firm (but not blind or spiteful) opposition. Each must run the gauntlet that alone can test it.

For the self-correcting structure works. There is delay and heartbreak often enough, but it works. However grim and slow the self questioning process of science may be (indeed, that the process exists at all is a matter of pride to scientists), science remains man's only self-correcting intellectual endeavor.

Isaac Asimov, 1977[7]

---

While some scientists are endoheretics, pseudoscientists are *exoheretics*, in that they attack scientific orthodoxy from outside the system. Typically, they have little training in, and often a flawed comprehension of, the topic they study. They also tend to present their ideas directly to the general public, most of whom have little understanding of science, instead of writing for scientists. By skipping the peer review system of science, it often is easy to publish exoheretical ideas. It is common for a well-written piece of pseudoscience, such as Velikovsky's *Worlds in Collision*, or one on a topic of interest to many in the public, such as weight loss programs or UFOs, to find a large audience even when the ideas contained in it are mostly insupportable on scientific grounds.

Science works because a theory without good evidence is rejected or at least considered unreliable until good evidence is presented to support it. On the other hand, strong evidence is not a requirement for acceptance of theories by pseudoscientists. While careful research has shown that astrology should be rejected, it remains popular because unskeptical people want it to work. Belief in UFOs is based largely on hundreds of eyewitness accounts, yet it is clear that all of the accounts could be mistaken, and there is no evidence for alien visits which do not have more earthly (and more likely) explanations. The overwhelming majority of parapsychological experiments show no evidence of psychic powers or can be shown to be seriously flawed in design, yet belief in psi by many is unaffected by such information. Most creation scientists find reasons, often tenuous, to reject any evidence contrary to their theory and embrace any evidence that supports it; the quality of the evidence is not important to them. Unconditional belief is unacceptable in science yet is the norm in pseudoscience. In this respect, pseudoscience is more like religion than science; it is based on faith rather than reliable empirical evidence. While faith usually is well served in religion, it is dangerous in pseudoscience.

---

Science's success stems in large part from its conservatism, its insistence on high standards of effectiveness. Quantum mechanics and general relativity were as new, as surprising, as anyone could ask for. But they were believed ultimately not because they imparted an intellectual thrill but because they were effective: they accurately predicted the outcome of experiments. Old theories are old for good reason. They are robust, flexible. They have an uncanny correspondence to reality. They may even be True.

John Horgan, 1996[8]

---

### THE NEED FOR CRITICAL THINKING

The acceptance of pseudoscience by so many people is evidence of a shortage of critical thinking, and such shortages have been typical throughout human history. Many people see mysterious phenomena (e.g., UFOs, ESP) as fascinating and easy solutions to problems (even if the easy solutions do not work, like "miracle diets"), and therefore people accept them unquestioningly. Often, they are willing to spend their money, risk their health, and make important life decisions (like taking advice from psychics or astrologers on marriage and having children) without any thought as to the reliability of the claims on which these phenomena are based. A lack of critical thinking makes it easy for people to mistake nonsense for sense.

Pseudoscience is widely accepted, at least in part, because people in general are not very good at doing what scientists are supposed to do, namely, separate the real from the unreal. Thomas Gilovich discusses several reasons why people commonly believe what is not true. People tend to draw conclusions from unrepresentative information, and even when there is enough unbiased information available, they often interpret it in ways that confirm what they want to find there.[10] For example, a few testimonials from patients who believe that they have been cured by some form of quack medicine is enough to convince many that the procedure (or drug or whatever) works, and they often have no interest in finding out about patients who did not benefit from it.

---

Both [circus promoter P.T.] Barnum and [writer] H. L. Mencken are said to have made the depressing observation that no one ever lost money by underestimating the intelligence of the American Public. The remark has worldwide application. But the lack is not intelligence, which is in plentiful supply; rather, the scarce commodity is systematic training in critical thinking.

Carl Sagan, 1979[9]

---

One critical thinking skill in short supply is the ability to judge evidence fairly. As Gilovich shows in the accompanying quote, many people use different criteria to evaluate evidence they want to accept and evidence they want to reject.

---

... we tend to use different criteria to evaluate propositions or conclusions we desire, and those we abhor. For propositions we want to believe, we ask only that the evidence not force us to believe otherwise—a rather easy standard to meet, given the equivocal nature of much information. For propositions we want to resist, however, we ask whether the evidence *compels* such a distasteful conclusion—a much more difficult standard to achieve. For desired conclusions, in other words, it is as if we ask ourselves, "*Can* I believe this?", but for unpalatable conclusions we ask, "*Must* I believe this?" The evidence required for affirmative answers to these two questions are enormously different. By framing questions in such ways, however, we can often believe what we prefer to believe, and satisfy ourselves that we have an objective basis for doing so.

Thomas Gilovich, 1991[11]

---

Another reason pseudoscience is so popular is that people tend to believe what they are told, even when it comes from unreliable sources. Secondhand stories are usually less accurate than the original, yet many people consider them equally reliable. Some stories are altered to make them shorter, some to make the point more obvious, some to add entertainment value, and some to deliberately deceive.[12] Writers discussing the Bermuda Triangle, for example, often neglect to inform readers that some of the missing planes and ships used to illustrate the phenomenon were far away from the region when they disappeared, and some writers withheld information to make disappearances more mysterious.[13] Many books and television reports of paranormal phenomena make little or no attempt to show alternative explanations, yet unskeptical people believe them unquestioningly.

## A SKEPTICAL ATTITUDE

Science can only be effective if approached skeptically, and pseudoscience should be approached with the same critical attitude. Evidence used to support any claim must be evaluated with respect to its reliability. For any idea to overthrow conventional scientific knowledge, which most pseudoscientists seek to do, the evidence should be quite strong.

---

Those of us who question the beliefs in ghosts, reincarnation, telepathy, clairvoyance, psychokinesis, dowsing, astral influences, magic, witchcraft, UFO-abductions, graphology, psychic surgery, homeopathy, psychoanalysis, and the like, call ourselves "skeptics." By so doing we wish to indicate that we adopt Descartes's famous methodological doubt. This is just initial distrust of extraordinary perceptions, thoughts, and reports. It is not that skeptics close their minds to strange events but that, before admitting that such events are real, they want to have them checked with new experiences or reasonings. Skeptics do not accept naively the first things they perceive or think; they are not gullible. Nor are they neophobic. They are just critical; they want to see evidence before believing.

Mario Bunge, 1991[14]

---

Another point made by many skeptics is that the supporting evidence for pseudoscientific claims should come from the proponents. Scientists cannot prove that ESP, or Bigfoot, or UFOs do *not* exist; it is up to the backers of such claims to present strong evidence that they do exist. Anecdotal evidence, fuzzy photographs, and badly designed experiments are not sufficient for the task. Only when good evidence has been presented (and good in the opinion of reasonable critics, not proponents), should wild claims be taken seriously by scientists and others. In addition, the more outrageous the claim, the stronger the evidence should be before well-established knowledge is revised. Science must be open to new ideas and the evaluation of new evidence, but within limits.

> Keeping an open mind is a virtue—but ... not so open your brains fall out. Of course we must be willing to change our minds when warranted by new evidence. But the evidence must be strong. Not all claims to knowledge have equal merit.
>
> Carl Sagan, 1995[15]

This discussion of how pseudoscience should be treated may be interpreted as a pompous attitude on the part of scientists with respect to claims made by nonscientists. Simply put, the institution of science has spent centuries developing the best, though definitely imperfect, way to learn about real phenomena, to test claims, and to avoid self-delusion. If claims about real things are made, as is done by pseudoscientists, they should not be trusted until subjected to scientific scrutiny. If the scientists with expertise in a particular phenomenon all agree that a claim is likely to be incorrect, then that collective opinion should carry considerable weight. It is far more likely that they are right than wrong, as discussed by Bertrand Russell in the opening quote to this chapter. On the other hand, it is the duty of scientists to make such judgments on the basis of the evidence and not for some financial, political, or religious reason or from a truly pompous attitude, such as "I'm a scientist and if nonscientists disagree with me they must be wrong." Scientists have a duty to the public to make informed and impartial judgments within their area of expertise. They also have a duty not to present themselves as scientific experts in areas for which they have no expertise.

Scientists also have a duty to act fairly. Some scientists attacking Velikovsky's ideas unintentionally added to his public support by appearing dogmatic about the incorrectness of his work and also attacking (verbally, at least) Velikovsky personally, instead of dealing only with his ideas, reasoning, and supporting evidence. Some scientists even boycotted the company that published his book.[16] Such behavior is unacceptable in any form of argument, and scientists damage the reputation of all science by acting in such ways. A reasoned discussion of ideas is the most effective way to confront pseudoscience, with scientific evidence to back up claims. For scientists to resort to disreputable practices is not in the best interest of science or the public.

### EVALUATING EXTRAORDINARY CLAIMS

Theodore Schick and Lewis Vaughn suggest a way for people to evaluate extraordinary claims, and one need not be a practicing scientist to follow their advice. They call it the SEARCH formula, which is an acronym for the steps in the approach: (1) State the claim, (2) examine the Evidence for the claim, (3) consider Alternative hypotheses, and (4) Rate, according to the Criteria of adequacy, each Hypotheses.[17]

When confronted with an extraordinary claim, the first step is to clarify it, or to *state the claim* clearly and specifically. Such vague statements as "Velikovsky was right" are difficult to evaluate. "A comet had a near collision with Earth in historical times" is amenable to testing and achieving a "probably yes" or "probably no" answer.[18]

The second step is to *examine the evidence for the claim*. Both the quality and the quantity of the evidence supporting the claim need to be evaluated. Is the evidence relevant to the problem? What are its limitations? Controlled experiments usually are better than anecdotal evidence, as discussed in Chapter 7. Is the evidence presented biased? Bias often is a problem, for example, with testimonials from patients cured by a procedure, while information on those not cured is not made available. Along the same lines, evidence provided by people without a financial or political stake in the phenomenon often merits consideration as more reliable than evidence from the people who stand to benefit from the claim.

In other words, the more objective the evidence, the more likely it is to be reliable.[19]

With extraordinary claims, it is best to *consider alternative hypotheses*. Thinking about your sister and then receiving a call from her can generate several hypotheses. "This is evidence of ESP" is one, but so are "it was purely coincidence" and "she often calls me when she gets home from work, which may explain why I was thinking of her at that time." Once a set of hypotheses has been identified, they can be compared according to the criteria of adequacy.

Schick and Vaughn presented their *criteria of adequacy* to ". . . help us determine how well a hypothesis (scientific, pseudoscientific or somewhere in between) accomplishes the goal of increasing our knowledge, and therefore the degree to which we should consider it an accurate theory."[20] The criteria are testability, fruitfulness, scope, simplicity, and conservatism.

As discussed in Chapters 2 and 3, scientific hypotheses must be *testable*. "More babies are born during days with a full moon than days with other phases of the moon," for example, is a testable hypothesis. Research can indicate if it is likely to be correct or not. "Paranormal phenomena exist, but go away when a skeptical person is present"[21] is not a testable hypothesis. Such a hypothesis allows negative results to be nullified, and therefore, any test is meaningless. Untestable hypotheses are useless in explaining real phenomena.

Just because a hypothesis is testable, though, does not mean that it is true. Good hypotheses also are *fruitful*, in that they explain more than they were originally intended to explain. Darwin's theory of evolution explained the shapes of beaks of finches on the Galapagos Islands, as discussed in Chapter 2, but it also explains much of the physiology and behavior of all species of plants and animals. It is very fruitful. "Other things being equal, hypotheses that make accurate, unexpected predictions are more likely to be true than hypotheses that don't."[22]

Along the same lines, the *scope* of a hypothesis is important. When two hypotheses explain a phenomenon equally well, the one which also explains other phenomena should be regarded more favorably. Albert Einstein's theory of relativity won out over Isaac Newton's theories of gravity and motion because Einstein's theory explained everything Newton's could, but it also explained more, like the orbit of Mercury, as discussed in Chapter 2.[23]

*Simplicity* is also a consideration when comparing hypotheses. Occam's razor, discussed in the previous chapter, is important here. If two hypotheses explain a phenomenon equally well, the simplest one—the one making the fewest assumptions—is more likely to be correct. Using the hypotheses for the phone call presented previously, the second, that it is coincidence, makes fewer assumptions than the one that involves ESP.

The last criteria of adequacy is *conservatism*. Well-established theories are trustworthy for a reason—usually, they are the best explanations for the phenomena in question. They may be wrong, but it is more likely that they are reasonably correct.[24] Perpetual motion machines are supposed to operate indefinitely without any input of energy, which violates the laws of thermodynamics. Since these physical laws very effectively describe how energy behaves, it is considered impossible for such devices to work, and the U.S. Patent Office, for example, generally refuses to consider patents for them.[25] While most scientific knowledge is not as completely verified as the laws of thermodynamics, theories accepted by most experts in a field should be considered more reliable than alternative explanations most of the time.

The criteria of adequacy are not a foolproof way to find the truth. All scientific hypotheses must be testable, but fruitfulness, scope, simplicity, and conservatism each merely suggest that one theory is more likely to be correct than another, but not with any certainty. In comparing hypotheses, the relative importance of the criteria depends on the specific situation.[26]

Nevertheless, the criteria of adequacy often are useful in separating pseudoscience from science. If a theory rates favorably on all counts, as evolution does, then people should consider it a reliable description of that part of reality. If a theory rates poorly on most of the criteria, as does creation "science," then it should not be accepted as accurate.[27] When obvious conclusions are difficult to reach using the criteria, it is best to suspend judgment until further research clarifies the situation. While much of the evidence supporting the reality of the Loch Ness Monster, for example, is of questionable value, the existence of the creature(s) should not be rejected or accepted by skeptical people.

All critical thinkers have a skeptical attitude. Ideally, science education helps students indirectly by pushing them to develop the critical thinking skills necessary to evaluate all kinds of phenomena, scientific, pseudoscientific, and other. This is well-expressed in the concluding quote, from Thomas Gilovich.

It stands to reason, then, that greater familiarity with the scientific enterprise should help to promote the habits of mind necessary to think clearly about evidence and to steer clear of dubious beliefs. Involvement in the process and concepts of science not only teaches these habits of mind directly, it also provides experience with problems, phenomena, and strategies from which they can sometimes be intuited, or at least more deeply understood. Also, one who participates in the scientific enterprise receives valuable exposure to uncertainty and doubt. Because science tries to stretch the limits of what is known, the scientist is constantly thrust against a barrier of ignorance. The more science one learns, the more one becomes aware of what is *not* known, and the provisional nature of much of what is. All of this contributes to a healthy skepticism toward claims about how things are or should be. This general intellectual outlook, this awareness of how hard it can be to really know something with certainty, while humbling, is an important side benefit of participating in the scientific enterprise.

Thomas Gilovich, 1991[28]

## PROBLEMS

1. What is the difference between science and pseudoscience? Is there a definite boundary between the two?

2. What is the difference between endoheretics and exoheretics?

3. "Science is supposed to be self-correcting." Discuss this statement.

4. Read the first boxed quote from Thomas Gilovich (starting with ". . . we tend to use . . ."). Give an example of different standards for evidence in what we want to believe and what we do not want to believe, either from you personally or from a public figure.

5. "Keeping an open mind is a virtue—but not so open your brain falls out." Discuss this line from the second boxed quote from Carl Sagan.

6. Discuss the opening quote for this chapter by Bertrand Russell.

7. Investigate one form of pseudoscience so that you have a good understanding of its main claim and the justification presented for accepting it. Apply the SEARCH formula to evaluate that claim. What conclusion do you reach?

8. Find two competing theories in a particular area of science, and learn about their claims. Apply the SEARCH formula to each. Does this exercise suggest that one of the theories is stronger than the other?

9. Discuss the ending quote by Thomas Gilovich. What is the connection between science education and critical thinking? How can science courses at all age levels be designed to teach critical thinking?

## NOTES

1. Russell, Bertrand, 1928, *Sceptical Essays*, London: George Allen and Unwin, Ltd., p. 12–13. See also Schick, Theodore, Jr., and Vaughn, Lewis, 1995, *How to Think about Weird Things: Critical Thinking for a New Age*, Mountain View, California: Mayfield Publishing Co., p. 106–9.

2. For general discussions of common themes in pseudoscience, see Bunge, Mario, 1984, "What Is Pseudoscience?" *Skeptical Inquirer* 9: 36–46 and Radner, Daisie, and Radner, Michael, 1982, *Science and Unreason*, Belmont, California: Wadsworth Publishing Co.

3. Asimov, Isaac, 1977, "Forward: The Role of the Heretic," in *Scientists Confront Velikovsky*, ed. Donald Goldsmith, Ithaca, New York: Cornell University Press, p. 7–15.

4. Sagan, Carl, 1979, Broca's Brain: Reflections on the Romance of Science, New York: Random House, p. 64.

5. Gould, Stephen Jay, 1977, *Ever Since Darwin: Reflections in Natural History*, New York: W. W. Norton, Inc., p. 154. (The quote is from Chapter 19, "Velikovsky in Collision," p. 153–59.)

6. Asimov, 1977. See also Strahler, Arthur N., 1992, *Understanding Science: An Introduction to Concepts and Issues*, Buffalo: Prometheus Books, p. 200–201.

7. Asimov, 1977, p. 10

8. Horgan, John, 1996, *The End of Science: Facing the Limits of Knowledge in the Twilight of the Scientific Age*, Reading, Massachusetts: Helix Books, p. 137.

9. Sagan, 1979, p. 49

10. Gilovich, Thomas, 1991, *How We Know What Isn't So: The Fallibility of Human Reason in Everyday Life*, New York: The Free Press, p. 29–72.

11. Ibid., p. 83–84; italics in original.

12. Ibid., p. 88–111.

13. Kusche, Lawrence David, 1995, *The Bermuda Triangle Mystery—Solved*, New York: Harper and Row, p. 276–77.

14. Bunge, Mario, 1991, "A Skeptic's Beliefs and Disbeliefs," *New Ideas in Psychology* 9: 131–49; quote is on p. 131.

15. Sagan, Carl, 1995, *The Demon-Haunted World: Science as a Candle in the Dark*, New York: Random House, p. 187. (Sagan attributes the line "keeping an open mind is a virtue—but not so open your brains fall out" to space engineer James Oberg.)

16. Bauer, Henry H., 1984, *Beyond Velikovsky: The History of a Public Controversy*, Urbana: University of Illinois Press, p. 354.

17. Schick and Vaughn, 1995, p. 235–84.

18. Ibid., p. 238.

19. Ibid., p. 238–39.

20. Ibid., p. 201.

21. Hines, Terrence, 1988, *Pseudoscience and the Paranormal: A Critical Examination of the Evidence*, Buffalo: Prometheus Books, p. 81–82.

22. Schick and Vaughn, 1995, p. 241.

23. Ibid., p. 206.

24. Ibid., p. 210.

25. Rothman, Milton A., 1988, *A Physicist's Guide to Skepticism*, Buffalo: Prometheus Books, p. 19–35, 141–42.

26. Schick and Vaughn, 1995, p. 210.

27. Ibid., p. 211–19.

28. Gilovich, 1991, p. 189; italics in original.

# CHAPTER 9

# THE PROCESS OF SCIENCE

SOCRATES: And the same object appears straight when looked at out of the water, and crooked when in the water; and the concave becomes convex, owing to the illusion about colours to which the sight is liable. Thus every sort of confusion is revealed within us; and this is that weakness of the human mind on which the art of conjuring and of deceiving by light and shadow and other ingenious devices imposes, having an effect upon us like magic.

GLAUCON: True.

SOCRATES: And the arts of measuring and numbering and weighing come to the rescue of the human understanding—there is the beauty of them—and the apparent greater or less, or more or heavier, no longer have the mastery over us, but give way before calculation and measure and weight?

GLAUCON: Most true.

SOCRATES: And this, surely, must be the work of the calculating and rational principle in the soul?

GLAUCON: To be sure.

SOCRATES: And when this principle measures and certifies that some things are equal, or that some are greater or less than others, there occurs an apparent contradiction?

GLAUCON: True.

SOCRATES: But were we not saying that such a contradiction is impossible—the same faculty cannot have contrary opinions at the same time about the same thing?

GLAUCON: Very true.

SOCRATES: Then that part of the soul which has an opinion contrary to measure is not the same with that which has an opinion in accordance with measure?

GLAUCON: True.

SOCRATES: And the better part of the soul is likely to be that which trusts to measure and calculation?

GLAUCON: Certainly.

SOCRATES: And that which is opposed to them is one of the inferior principles of the soul?

GLAUCON: No doubt.

*Plato, Fifth Century B C* [1]

The first eight chapters have provided an outline of various aspects of science, both what it is and what it is not. Building on that information, this chapter gives a brief and simplified overview of the entire process of science. In other words, how is it that we gain scientific knowledge? Science is not a linear process; the path taken from new idea to established knowledge typically is circuitous. The accompanying figure, *The Scientific Knowledge Acquisition Web* (SKAW), is a summary of how science is done.

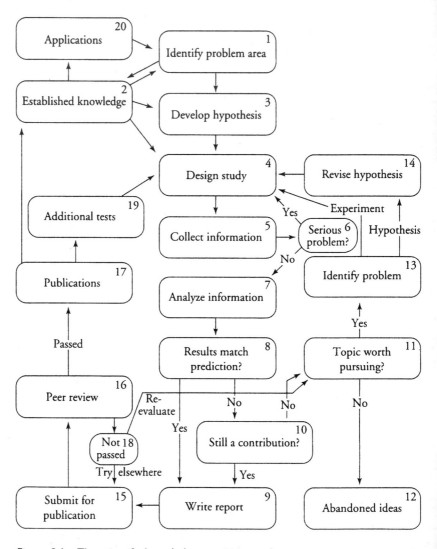

Figure 9.1    The scientific knowledge acquisition web.

## THE SCIENTIFIC KNOWLEDGE ACQUISITION WEB

There is no obvious starting point in SKAW because each part is connected to others. However, identifying a problem area (Box 1 in SKAW) is as good a place to start as any. Scientists see things they cannot readily explain in the course of their work. (Why are there so few birds around here? Why did I get such strange results in the lab last week?) They may read about problem areas in a research field. (The weakest link in our understanding of ____ is ____, and clearly this is an area of needed research.) They may have someone offer to pay for research on a given topic. (Will college students buy our product? What's the best way to dispose of our toxic waste? We will pay for research on improving military weapons systems.) With a general topic in mind, the scientist then learns what is known about it (Box 2) in preparation for his or her own research. (Some advocate not knowing much of the conventional knowledge of a topic before starting a research project in order to bring fresh ideas unbiased by others' opinions. In this approach, learning of other findings comes later.[2])

The next step in the process is to generate a hypothesis (Box 3). A potential explanation is stated and used to make a testable prediction of what would be found under specific experimental conditions. (When all other variables are held constant except temperature, I expect to get results that show ____.)

The hypothesis needs to be tested, and the design of the experiment is crucial (Box 4). The experiment is planned so that a clear answer can be obtained showing that the hypothesis is supported or not supported by the data. Part of preparing for a study is the logistical aspects, which can include getting funding for the project, assembling a team of assistants, procuring the needed equipment, and so on. This part of the process, especially obtaining funding, can take up a sizable fraction of a scientist's time.

The experiment is conducted (Box 5) by following the design as closely as possible to collect the required information, generally in the form of measurements of one form or another. Experiments rarely go exactly as planned, and typically, changes are made in the design when unexpected problems arise (Box 6). If the problem is minor, then the experiment usually proceeds in a modified form. If the problem encountered is serious, then the experiment may need to be redesigned and started over (Box 4).

The data collected in an experiment are then analyzed (Box 7). Raw data, the numbers collected as part of the experiment, are assembled and manipulated in order to make sense of them.

When the analysis is done, the results are compared to the prediction made by the hypothesis (Box 8). If the results do match the prediction, then the study is written up for publication (Box 9). If they do not match or they are inconclusive, then the question is asked "Is the study still a contribution to knowledge?" (Box 10). If the answer is "Yes," meaning that others can learn useful things from the study, then it is written up (Box 9); negative results can be as informative as positive ones to people trying to understand something in science. If, however, the answer is "No," the study by itself is not a contribution, then the scientist must decide if the topic is worth pursuing (Box 11). In some cases, the results of the study will teach the scientist that progress on the topic is unlikely, and it is time to abandon it; her or his time can be better used on other subjects (Box 12). If the project is worth pursuing, then the experience gained will aid in the preparation of another study on the same topic, which is more likely to generate useful results.

If the topic is worth pursuing, the problem or problems with the previous study must be identified (Box 13). If the problem was most likely with the hypothesis, meaning that it was not a good explanation of the problem, then a new or modified hypothesis is created (Box 14) and a new study is prepared (Box 4). If the problem was more likely with the experiment instead of the hypothesis, then a modified experiment is planned to get around the problem (Box 4 again). Whether the problem is with the hypothesis or the experiment, a new experiment is designed and the process continues.

Eventually, the study will be seen by the scientist as a significant contribution to knowledge (Box 10), unless it has been abandoned (Box 12). A study that is deemed worthy to share with the scientific community is written up for publication and submitted to a journal (back to Box 9). The paper generally consists of an overview of what is known on the topic, the methods used in the study, the results, and a discussion of how the results fit in with established knowledge on the topic. (These results support the theory that ___. These results suggest that our generally held theory needs to be modified to include ___.)

The paper is then submitted for publication, usually to a scientific journal (Box 15). The journal editor asks other experts on the topic to evaluate the manuscript in terms of the significance of the topic, appropriateness of the methods used, and other criteria (Box 16). If the peer reviews are favorable, then the editor accepts the paper and publishes it (Box 17), often after revisions are made. If the paper is rejected by the editor (Box 18), then the scientist either submits it as is or in a revised form to another journal (going through Box 15 again), hoping for more favorable reviews this time, or asks himself whether or not the topic is worth pursuing (going back to Box 11).

Publication of a study puts it out into the realm of "established knowledge" where it is available to anyone interested in the topic (Box 2). "Established knowledge" here does not mean "True," it just indicates that the conclusions are backed up by one or more studies, and it can contribute to the development of theories in that particular area. Over time, some theories gain acceptance, and others are rejected by the community of scientists in that field.

---

But I believe that there is no philosophical highroad in science, with epistemological signposts. No, we are in a jungle and find our way by trial and error, building our road *behind* us as we proceed. We do not *find* signposts at crossroads, but our own scouts erect them, to help the rest.

Max Born, 1943[3]

---

Established knowledge affects the development of new problem areas (Box 1), new hypotheses (Box 3), and new ideas in experimental design (Box 4). Other scientists reading the paper may become intrigued by the findings and start exploring the problem area, some already involved in similar research may develop new hypotheses from the study, and still others may learn how to do better experiments.

The scientist who conducted the original work will have gained experience in studying the topic and may choose to continue with more research (Box 19). She or he may choose to refine the test to provide more convincing evidence of the value of the hypothesis or, in a similar vein, find other ways to test the hypothesis. Other scientists may begin working on the topic, learning from experience gained in the earlier studies. They will find new ways to test the hypothesis as well. The more carefully a hypothesis is tested and the number of different tests it has passed all contribute to the confidence that scientists place in its accuracy.

The development of theories leads, in some cases, to the application of that knowledge in the invention or improvement of technologies (Box 20). New or improved devices and new or revised government or business policies can all come from the application of scientific knowledge. Conversely, the perceived need for a new technology can drive the scientific process by showing the need for new knowledge in a particular area (Box 1). For example, basic research on genes led to the development of genetic engineering, and the use of genetic engineering has shown which aspects of genetic research should be emphasized to improve the knowledge needed to advance the further engineering of genes.

Nonexperimental scientific research is similar to experimental research in terms of the process by which knowledge is gained. The main difference is that hypotheses are not involved. After identifying a problem area (Box 1), a study is designed to provide information that will shed light on that problem (Box 4) and the information is collected (Box 5). Since there is no prediction with which to compare the results, the analysis (Box 7) is followed by the question "Is it a contribution?" (Box 10). From that point on, the process is the same as for experimental research.

Some scientists are involved in all aspects of SKAW, while others concentrate on certain parts. Albert Einstein, for example, is best known for his work in identifying problem areas and developing hypotheses. He generally left the testing of his ideas to others. Other scientists work primarily in refining experimental techniques or analysis procedures. Some spend most of their time editing journals or working in funding agencies. Scientists may work alone or in teams ranging in size from two to hundreds. How a scientist contributes depends on his or her personality, strengths, and opportunities.

CONCLUSIONS

Chapter 3 opened with John Godfrey Saxe's poem of the "Blind Men and the Elephant." These fictitious men could have learned much about the elephant if they had understood and followed the process of science as outlined here.

The Blind Men and the Elephant, Part 2
(with apologies to John Godfrey Saxe)

I.

They talked, those men from Indostan
  While standing at the door,
Of elephants and how they look
  (This talk was such a bore!),
At last agreed that knowledge gained
  Requires something more.

II.

Perhaps each one, in his own way,
  Did learn about a bit
Of the beast's elusive mystery,
  But just a part of it
With work, they thought, they just
    might see
  The puzzle pieces fit.

III.

'Twas obvious to all of them
  For learning to progress,
That they must share in what they found—
  Jointly sort out the mess.
And seek to fully understand
  Elephants, more or less.

IV.

Some worked alone and some in teams,
  In both the field and lab.
Models were made: some soft, some hard
  Some good, some pretty bad.
But when they pooled the useful work,
  At truth they made a grab.

V.

They checked each other's methods out,
  Some kept, some put asunder.
To use the ones which passed the test
  Reduced the chance of blunder.
Then they'd trust what they had
    learned
  Of elephants' fine wonders.

VI.

They made great strides in what they
    knew
  Of the nature of the beast.
Of what and where and how and why
  They knew much more at least.
For blind men learned how best to learn
and vision soon increased!

This endeavor we call science is one of the crowning achievements of humanity. People have an uncanny ability to assume that what we think is true is true and that what we want to believe is true is true. The beauty of science is the realization that our images and desires are not the same as reality and that we have set up a way to understand the universe on its own terms. This is tremendously difficult to do, and we probably have never succeeded completely in any area of science. The fact that we try is truly amazing.

## PROBLEMS

1. Write a short story on the path a scientific idea takes as it travels through SKAW. Have fun with it, but show your knowledge of the process of science.

2. If you were a scientist, in which areas of SKAW would you prefer to work? Or would you prefer to work in all areas? Why?

3. Explain the quote from Max Born.

4. What is the connection between the opening quote, from Plato's *Republic*, and the final paragraph of this chapter?

## NOTES

1. Plato, *The Republic,* Book 10, trans. Benjamin Jowett, Cleveland: The World Publishing Company, 1946, p. 360–61.

2. See, for example, Root-Bernstein, Robert Scott, 1989, *Discovering*, Cambridge: Harvard University Press, p. 54.

3. Born, Max, 1943, *Experiment and Theory in Physics*, Cambridge: Cambridge University Press, p. 44; italics in original.

# UNITS

... men used an incredible diversity of units, derived from unknown sources and forgotten accidents. A "foot" might equal 10 modern inches, or 13, or 17, or even 27. Half a dozen versions of the foot could be found in such a city as Rome, and a dozen other units of length with equally uncertain values. It depended on what was being measured, and when, and by whom, and whether one was buying or selling. A merchant used one measure when he bought, another—identically named—when he sold.

This is history, yet not all is ancient or medieval. In our own century, in Brooklyn, New York, there was a time when the city surveyors recognized as legal four different "feet": the United States foot, the Bushwick foot, the Williamsburg foot, and the foot of the 26th Ward. All legal, all different. Some strips of Brooklyn real estate were untaxable, because, after two surveys, made with different units, these strips, legally, didn't exist!

*John Perry, 1955*[1]

Standardization of the units of measurement has greatly helped commercial transactions, taxation, Brooklyn real estate, and science. The latter subject is the concern of this appendix. Other topics useful for scientists to know will be discussed in the other three appendices.

Measurement is a fundamental part of science; it helps narrow the difference between human perceptions of the real world and the real world itself. Any measurement has two parts: the value and the unit. To write that two trees are 25 apart is meaningless because 25 has no unit. To say that the trees are 25 *meters* apart is informative. Because units are fundamental to scientific communication, rules for the use of units have been established. In science, the *metric system* is used. Actually, a particular version of the metric system is used, known as the *International System of Units*, abbreviated *SI* (for Le Système International d'Unités). The purpose of this appendix is to introduce SI, but not to elaborate in detail on individual units or on how they are used in various scientific disciplines.

> [T]he United States,...with Burma and Liberia, is still holding out against the metric system....Despite an act signed by President Ronald Reagan in 1988 that specifies the metric system as the preferred system of measurement for trade and commerce in the United States, road signs are still in miles, people still buy fish by the pound and swelter in 100-degree Fahrenheit heat.
>
> Malcolm W. Browne, 1993[2]

The metric system was developed in France. Following the French Revolution in 1789, the government set out to establish a coherent system of units based on properties of natural phenomena. It was to be a decimal system, meaning based on powers of ten. Almost all countries of the world use the metric system for commerce and industry, and all scientists are expected to use SI. The International Committee for Weights and Measures is located near Paris and oversees the details of SI; they modify the system occasionally.

As an international endeavor, science should be communicated in such a way that scientists from a wide variety of countries and cultures can understand each other's findings. SI provides a useful standard by which measurements are made, and it also provides general rules for reporting the measurements. If SI is used properly in science, much confusion and wasted effort at unit conversions and interpretations can be avoided.

### THE INTERNATIONAL SYSTEM OF UNITS

In SI, there is only one official unit for any physical quantity. Units are divided into two classes: base units and derived units. There are seven base units, and all derived units are created by combining base units.

#### Base Units

The seven SI base units are meter, kilogram, second, kelvin, ampere, mole, and candela. They are shown in Table A1.1, and their official definitions are given in Table A1.2. To keep this description of base units relatively simple, each one will be described briefly, but the definitions will not all be fully explained.[3]

Table A1.1   SI base units

| QUANTITY | NAME | SYMBOL |
| --- | --- | --- |
| length | meter | m |
| mass | kilogram | kg |
| time | second | s |
| temperature | kelvin | K |
| electric current | ampere | A |
| amount of substance | mole | mol |
| luminous intensity | candela | cd |

The unit of length is the *meter*. Originally, the meter was defined as one ten-millionth of the distance from the equator to the North Pole. Careful surveying from Barcelona, Spain, to Dunkirk, France, in 1792–98 resulted in an estimate of the total length from the equator to the Pole, and a metal bar was constructed to be used as the standard reference for the meter. The metal is an alloy of platinum and iridium, chosen for its stability. It has since been determined that the meter as originally defined is not one ten-millionth of the distance from the equator to the North Pole (that distance is closer to 10 002 290 meters), but the length of a meter has not been adjusted to account for this discrepancy.[4] However, the definition of the meter has changed. The metal bar is no longer the standard reference. The meter is now defined as the distance light travels in 1/299 792 458 of a second, as shown in Table A1.2. A metal bar may be lost or damaged, but the speed of light in a vacuum will always be the same, and the meter can be determined experimentally by anyone with the right equipment and the knowledge to use it properly.

The unit of mass is the *kilogram*. Originally, the kilogram was defined as the mass of water in a cube one-tenth of a meter on a side. A reference kilogram was constructed, also of platinum and iridium, with a mass equal to such a cube of water to serve as the international standard. Unlike the reference meter bar, no alternative definition for mass has

Table A1.2    Official definitions of base units[5]

| | |
|---|---|
| METER: | The meter is the length of the path travelled by light in vacuum during a time interval of 1/299 792 458 of a second. |
| KILOGRAM: | The kilogram is the unit of mass; it is equal to the mass of the international prototype kilogram. |
| SECOND: | The second is the duration of 9 192 631 770 periods of the radiation corresponding to the transition between the two hyperfine levels of the ground state of the cesium-133 atom. |
| AMPERE: | The ampere is that constant current which, if maintained in two straight parallel conductors of infinite length, of negligible circular cross section, and placed 1 meter apart in vacuum, would produce between these conductors a force equal to $2 \times 10^{-7}$ newton per meter of length. |
| KELVIN: | The kelvin, unit of thermodynamic temperature, is the fraction 1/273.16 of the thermodynamic temperature of the triple point of water. |
| MOLE: | 1. The mole is the amount of substance of a system which contains as many elementary entities as there are atoms in 0.012 kilogram of carbon 12. 2. When the mole is used, the elementary entities must be specified and may be atoms, molecules, ions, electrons, other particles, or specified groups of such particles. |
| CANDELA: | The candela is the luminous intensity, in a given direction, of a source that emits monochromatic radiation of frequency 540 $\times 10^{12}$ hertz and that has a radiant intensity in that direction of (1/673) watt per steradian. |

been made, so the prototype kilogram remains the reference mass. Advances in the measurement of atomic masses likely will lead to the kilogram being redefined and possibly renamed in the future according to an atomic-based mass standard.[6]

The *second* is the SI unit of time. Originally, the second was defined in terms of the earth's rotation; each second is 1/86 400 of the length of a day (24 hours times 60 minutes times 60 seconds equals 86 400).

Because the speed of the rotation of the earth is not constant, the definition of a second has changed and involves the frequency of radiation given off by cesium-133 atoms under special conditions. (The official definition is too complicated for this discussion.)

The unit of temperature is *kelvin*. The zero point for kelvin is *absolute zero*, or the lowest temperature theoretically possible. Kelvin is related to the more commonly used *degree Celsius*, such that 0°C equals 273.15 K. In the Celsius scale, 0°C is the approximate temperature at which water freezes, and 100°C is the approximate temperature at which water boils at average sea level air pressure.

The *ampere* is the SI unit of electric current. It is the amount of current required to maintain a given force between two conductors, under specific conditions. (Like the definition of the second, the definition of the ampere is too complicated for this book.)

In chemistry, there is a unit called the *mole*, which is the amount of substance. The mole is the number of "things" (atoms, electrons, molecules, etc.) as there are atoms in 0.012 kilograms of carbon 12. The number of atoms in 0.012 kilogram of carbon 12 is not known precisely (it is about $6.02 \times 10^{23}$), though the mass of a "thing" relative to that of the carbon-12 atom can be determined.

The *candela* is the unit of luminous intensity, which can be thought of as "brightness." Originally, luminous intensity was defined in terms of the amount of light given off by candles. It is now defined in a complicated way by the amount of radiation given off by an object under specific conditions.

### Derived Units

All other units in SI are called *derived units* and are created using the base units. Table A1.3 lists some SI derived units. For example, speed, which is length divided by time, has the SI unit of meters per second. Area is length squared, so the SI unit is meter squared. Density is mass contained in a given volume, so the SI unit of density is kilograms per meter cubed. Some derived units are named from the base units used, such as meters per second, while others have been given special names, such as newton for force and pascal for pressure. Keep in mind, though, that SI base units are used to determine newtons and pascals.

Table A1.3    Sample SI derived units

| QUANTITY | NAME | SYMBOL |
| --- | --- | --- |
| area | meter squared | $m^2$ |
| volume | meter cubed | $m^3$ |
| speed, velocity | meter per second | $m\ s^{-1}$ |
| acceleration | meter per second squared | $m\ s^{-2}$ |
| density | kilogram per meter cubed | $kg\ m^{-3}$ |
| plane angle | radian[a] | rad |
| solid angle | steradian[a] | sr |
| angular velocity | radian per second | $rad\ s^{-1}$ |
| frequency | hertz | Hz |
| force | newton | N |
| pressure, stress | pascal | Pa |
| energy, work | joule | J |
| power | watt | W |
| Celsius temperature | degree Celsius | °C |

[a] Radian and steradian are dimensionless derived units: radian = $m\ m^{-1}$ = 1  and steradian = $m^2\ m^{-2}$ = 1

The SI unit for two-dimensional angles is the radian. Until 1995, the radian, along with the *steradian* for three-dimensional angles, was given a special class, known as supplementary units. Now they are grouped with the derived units.[7] The *radian* is used for plane angles; think of angles on a circle. One radian is the angle formed by a wedge of the circle, where the arc of the circle inside the wedge is equal in length to the radius of the circle. In SI, radians are used for plane angles instead of degrees because radians can be determined with the base unit of meter (length of radius and length of arc), while degrees are unrelated to any base unit. (Because the steradian is used much less frequently than the radian, it will not be described here.) Radians and steradians can be used to make other derived units. Angular velocity, for example, is the rate at which something rotates, so the SI unit is radians per second.

## Prefixes

Prefixes are used when the sizes of the base units are not convenient for a given purpose. For example, it is easier to discuss the distance from Toronto to Vancouver in kilometers than in meters. SI prefixes are given in Table A1.4. One source of confusion is that when describing mass, kilogram is the base unit, though the prefixes are applied to gram (like milligram). This is due to a disagreement between two committees in the 1790s attempting to determine the unit of mass. One wanted the gram (the mass of a centimeter cubed of water) and the other wanted the "grave" (the mass of a decimeter cubed of water). The grave did become the official unit but was renamed the kilogram, possibly to appease the other committee.[8]

## Other Units Used with SI

There are certain units which are not part of SI but are acceptable because they are used widely. They are listed in Table A1.5. In some scientific discussions, the second is an impractical unit of time. For example, when discussing the time span of the Bronze Age, use of seconds is awkward; years is a more informative unit (the symbol for year is a, for annum). The speed of a vehicle may be given in kilometers per hour rather than the official meters per second. Liter (the volume of a decimeter cubed) should be used only for measures of fluids, dry measures, and volumetric capacity. Milli and micro are the only prefixes that should be used with liter. Degrees, minutes, and seconds are acceptable in map making and astronomy, but radian should be used for most other applications of plane angle. If degree is used in applications other than map making, fractions of degree are preferable to minutes and seconds. The term *metric ton* (which is 1000 kilograms) should not be used in science; megagram is the proper term. Hectare (an abbreviation of hectometer squared) and kilometer squared, should be used only for large land and water areas. These non-SI units should not be used to form other derived units.[9]

## Units Temporarily in Use

Certain units are commonly used in some countries, and it is acceptable for these units to be used until further notice. These include nautical mile

Table A1.4   Prefixes used with SI units

| MULTIPLICATION FACTOR | SCIENTIFIC NOTATION | NAME | ABBREV. |
|---|---|---|---|
| 0.1 | $10^{-1}$ | deci | d |
| 0.01 | $10^{-2}$ | centi | c |
| 0.001 | $10^{-3}$ | milli | m |
| 0.000 001 | $10^{-6}$ | micro | $\mu$ |
| 0.000 000 001 | $10^{-9}$ | nano | n |
| 0.000 000 000 001 | $10^{-12}$ | pico | p |
| 0.000 000 000 000 001 | $10^{-15}$ | femto | f |
| 0.000 000 000 000 000 001 | $10^{-18}$ | atto | a |
| 0.000 000 000 000 000 000 001 | $10^{-21}$ | zepto | z |
| 0.000 000 000 000 000 000 000 001 | $10^{-24}$ | yocto | y |
| 10 | $10^{1}$ | deka | da |
| 100 | $10^{2}$ | hecto | h |
| 1000 | $10^{3}$ | kilo | k |
| 1 000 000 | $10^{6}$ | mega | M |
| 1 000 000 000 | $10^{9}$ | giga | G |
| 1 000 000 000 000 | $10^{12}$ | tera | T |
| 1 000 000 000 000 000 | $10^{15}$ | peta | P |
| 1 000 000 000 000 000 000 | $10^{18}$ | exa | E |
| 1 000 000 000 000 000 000 000 | $10^{21}$ | zetta | Z |
| 1 000 000 000 000 000 000 000 000 | $10^{24}$ | yotta | Y |

(for length), knot (for speed), and bar (for pressure). A scientist should have a good reason for using these units, rather than their SI equivalents, though.

### Obsolete Systems and Terms

There are unit systems other than SI which have been used in science but are now obsolete. These include U.S. customary units (e.g., pounds, feet, degree Fahrenheit) and the CGS units (units based on centimeters, grams, and seconds rather than the meter, kilogram, and second basis of

Table A1.5   Units used with SI

| NAME | SYMBOL | VALUE IN SI UNITS |
|---|---|---|
| minute | min | 1 min = 60 s |
| hour | h | 1 h = 3600 s |
| day | d | 1 d = 86 400 s[a] |
| year | a | 1 a = 31 557 600 s[a] |
| degree | ° | $1° = (\pi/180)$ rad |
| minute | ' | $1' = (\pi/10\ 800)$ rad |
| second | " | $1" = (\pi/648\ 000)$ rad |
| liter | l, L[b] | $1\ L = 10^{-3}\ m^3$ |
| metric ton[c] | t | $1\ t = 10^3\ kg = 1\ Mg$ |
| hectare | ha | $1\ ha = 10^4\ m^2$ |

[a] As the speed of Earth's rotation slowly changes, these are approximate values.

[b] Both symbols for liter (lower- and uppercase L) are officially acceptable in SI, though L is preferred because it is less likely to be confused with the number one.

[c] Metric ton may be used in commercial but not scientific practice.

SI). Several terms used for units are obsolete and should not be used. These include degrees centigrade (equivalent to degree Celsius) and micron (an abreviation for micrometer).[10]

## RULES FOR WRITING UNIT SYMBOLS, PREFIXES, AND NUMBERS

When writing SI unit symbols, prefixes, and numbers, certain rules should be followed.[11] The symbols are written in lowercase letters, except when the unit is named for a person, like watt (symbol W) and newton (symbol N). Italic type is not to be used for symbols. Do not add an *s* to make a unit plural (meter and meters are both symbolized by m), and symbols are not to be followed by a period, except at the end of a sentence, of course. The multiplication of two or more units may be shown either with a raised dot or with a blank space between them. For example, a joule (which is a newton multiplied by a meter) may also be written as

N m   and   N·m

For division of units, a negative exponent, a horizontal line, or a solidus (oblique stroke, /) may be used. For example, a pascal (a newton divided by a meter squared) may also be written as

$$N\,m^{-2} \quad \text{and} \quad \frac{N}{m^2} \quad \text{and} \quad N/m^2$$

To avoid confusion, the solidus should not be used twice on the same line unless parentheses are used to make the arithmetic unambiguous. Acceleration, for example, may be written as $m\,s^{-2}$, but not as m/s/s.

For prefixes, no space should be left between the prefix and the unit symbol, as in mm for millimeter. When the prefix is attached to the unit symbol, a new symbol has been created. For example, $mm^2$ is one millimeter squared, not one-thousandth of a meter squared.

A blank space should be placed between a number and its unit. For example, write $5\,m\,s^{-1}$, not $5m\,s^{-1}$. An exception to this rule is when using the degree, minute, and second symbols, as in latitude and longitude designations; for example, 101° 52' 26" W.

When long numbers (more than four digits on one or both sides of the decimal point) are written, separate the digits into groups of three, with a space in between, not a comma. In some countries, the comma is used as the decimal sign, while in others, a period is used. For example, 98 765.432 10 is correctly written, while 98,765.43210 is not. For numbers less than one, place a zero on the left side of the decimal point, as in 0.123. The terms *billion* and *trillion* should be avoided as they have differing meanings in different countries. In the United States, for example, a billion is a thousand million ($10^9$), but in many other countries, a billion is a million million ($10^{12}$), with *milliard* for $10^6$.

There are no official international spelling rules for SI; each country decides independently. "Meter," "liter," and "deka" have different spellings in the United States than in most other nations, where the French spellings of "metre," "litre," and "deca" are used.

The decisions regarding SI and its use are dominated by representatives from the physical sciences, though all scientists are expected to use SI when possible. This is reasonable in the biological and earth sciences, but the social sciences use many units not included in SI. Social scientists should follow SI practice, when appropriate, such as in discussing distances or energy. When no SI unit exists, such as for money, a scientist should determine which unit is conventionally used in a discipline and use it unless he or she has a good reason to use another.

## THE GREEK ALPHABET

Letters from the Greek alphabet are used as symbols in science. In SI, the symbol for micro is $\mu$ (lowercase mu), as shown in Table A1.4. Greek letters are also used for physical phenomena, such as $\gamma$ rays (gamma rays) and $\lambda$, the symbol for wavelength. To the uninitiated, these symbols can be confusing. For your convenience the Greek alphabet is presented here:

### The Greek Alphabet (uppercase followed by lowercase)

| | | | | | | | |
|---|---|---|---|---|---|---|---|
| A α | Alpha | Ι ι | Iota | | P ρ | Rho |
| B β | Beta | K κ | Kappa | | Σ σ | Sigma |
| Γ γ | Gamma | Λ λ | Lambda | | T τ | Tau |
| Δ δ | Delta | M μ | Mu | | Υ υ | Upsilon |
| E ε | Epsilon | N ν | Nu | | Φ φ | Phi |
| Z ζ | Zeta | Ξ ξ | Xi | | X χ | Chi |
| H η | Eta | O o | Omicron | | Ψ ψ | Psi |
| Θ θ | Theta | Π π | Pi | | Ω ω | Omega |

## NOTES

1. Perry, John, 1955, *The Story of Standards*, New York: Funk and Wagnalls Company, p. 4–5.

2. Browne, Malcolm W., 1993, "Yardsticks Almost Vanish as Science Seeks Precision," *New York Times*, 23 August 1993.

3. More detailed explanations of the definitions can be found in Browning, Dennis R., 1994, *Metric in Minutes: The Comprehensive Resource for Learning and Teaching the Metric System (SI)*, Belmont, California: Professional Publications. Many chemistry and physics textbooks provide more detail as well.

4. Nelson, R. A. 1981. "Foundations of the International System of Units (SI)." *The Physics Teacher* 19: 596–613.

5. Taylor, Barry N., 1991a, *The International System of Units (SI)*, NIST Special Publication 330 (1991) ed: 3–5.

6. Quinn, Terry J., 1991, "The Kilogram: The Present State of Our Knowledge," *IEEE Transactions on Instrumentation and Measurement* 40: 81–85. See also Taylor, Barry N., 1991b, "The Possible Role of the Fundamental Constants in Replacing the Kilogram," *IEEE Transactions on Instrumentation and Measurement* 40: 86–91 and DiFilippo, Frank; Natarajan, Vasandt; Boyce, Kevin R.; and Pritchard, David E.; 1994, "Accurate Atomic Masses for Fundamental Metrology," *Physical Review Letters* 73: 1481–84.

7. Quinn, T. J., 1996, "News from the BIPM," *Metrologia* 33, no. 1: 81–89.

8. Lee, Jeffrey A., 1995, "The International System of Units and Its Use in Geography and Related Disciplines," *Journal of Geography* 94: 592–98.

9. See the following for more detail on other units used with SI: A.S.T.M., 1992, *Standard Practice for Use of the International System of Units (SI) (the Modernized Metric System)*, Philadelphia: American Society of Testing and Materials, Designation E-380-92; A.N.S.I./I.E.E.E., 1992, *American National Standard for Metric Practice*, ANSI/IEEE Std. 268-1992, New York: American National Standards Institute and Institute of Electrical and Electronics Engineers; Taylor, Barry N., ed., 1991a, *The International System of Units (SI)*. National Institute of Standards and Technology Special Publication 330, Washington D.C.: U.S. Government Printing Office; Taylor, Barry N., 1995, *Guide for the Use of the International System of Units (SI)*, National Institute of Standards and Technology Special Publication 811, Washington D.C., U.S. Government Printing Office.

10. Taylor, 1995, p. 11

11. For more information on the rules of writing SI, see the following: A.S.T.M., 1992; A.N.S.I./I.E.E.E., 1992; Taylor, 1991a; Taylor, 1995; Browning, 1994; A.N.M.C., 1993, *Metric Editorial Guide*, 5th ed. (rev.), Bethesda, Maryland: American National Metric Council.

# WORKING WITH
# MEASURED VALUES

Consider a precise number that is well known to generations of parents and doctors: the normal human body temperature of 98.6° Fahrenheit. Recent investigations involving millions of measurements have revealed that this number is wrong; normal human body temperature is actually 98.2° Fahrenheit. The fault, however, lies not with Dr. Wunderlich's original measurements—they were averaged and sensibly rounded to the nearest degree: 37° Celsius. When this temperature was converted to Fahrenheit, however, the rounding was forgotten, and 98.6 was taken to be accurate to the nearest tenth of a degree. Had the original interval between 36.5° Celsius and 37.5° Celsius been translated, the equivalent Fahrenheit temperatures would have ranged from 97.7° to 99.5°. Apparently, dyscalculia can even cause fevers.

*John Allen Paulos, 1995*[1]

Rules exist not only for the proper use and presentation of units, but also for other aspects of scientific data. In this appendix, appropriate ways of presenting and some aspects of the mathematical manipulation of data are discussed. Included are the topics of significant digits, rounding, scientific and prefix notation, and how units are handled in calculations.[2] All scientists are expected to understand the rules presented here and to follow them.

## SIGNIFICANT DIGITS

No measurement in science is perfect; they all have error, but the amount of error is important. In reporting measurements, therefore, the precision of the measurement is important and should be properly conveyed to the reader. "The plants are 3 m apart" does not mean that they are exactly 3 m apart; it means that the plants are somewhere between 2.5 and 3.5 m apart. If you write "The plants are 2.95 m apart," you imply that you have measured the distance and found it to be between 2.945 and 2.955 m.

While the precision of a measurement is the topic of discussion here, it must be noted that precision and accuracy are not the same thing. If you measure the distance between two plants using a meter stick, you can be precise to within a centimeter or maybe even a millimeter. If you are careless and do not put the start of the meter stick correctly by the first plant, then you do not have an accurate measurement. Malfunctioning equipment typically affects the accuracy of the measurements, not the precision, and it is dangerous to assume that precise measurements are also accurate.

Back to precision. The digits that represent the precision of actual measurements are called *significant digits*. When a quantity is written properly, all digits are significant except for some zeros. Zeros before a digit are not significant; on 0.09 kg, for example, the two zeros are there only to show where the decimal point is and are, therefore, not significant. Zeros that are not place holders, however, are significant: 0.090 has two significant digits (the 9 and the last 0). Zeros at the end of a number, but before the decimal point, are usually not significant; "200 years ago" usually does not mean exactly 200 years ago, it means about 200 years ago. "201 years ago," on the other hand, means 201 years (give or take half a year). "200.0 m" suggests a precise measurement to within plus or minus 0.05 m. The number of significant digits indicates the precision of a measurement. 1234.56 kg has six significant digits, 123.4 kg has four significant digits, 0.01 m has one, and, 0.003 m also has one. A measurement of 300 s has one significant digit, while 300.0 s has four. (Actually, 300 s may have one, two, or three significant digits, depending on the precision of the measurements. In such a case, the number of significant digits should be made clear, as in 300 s (± 0.5 s), if that is the precision of the measurement. Assume the fewest significant digits unless you know that the precision is greater than that.) Often the precision of a measurement is known, but sometimes it must be estimated. If you carefully measure distance with a meter stick, you know that the precision is to the millimeter, for example. If, however, you are using someone else's measurements, you may have to estimate the precision, based on the methods used.

Measured quantities always have significant digits. Other numbers may have perfect precision. "Five people were interviewed" means exactly five people, not 4.5 to 5.5 people. By definition, one meter is exactly 1000 millimeters, not about 1000. Significant digits do not apply to such exact numbers.

## ROUNDING

Quantities are rounded to bring them to the proper number of significant digits, after either converting units or using quantities in mathematical operations. It is improper to report quantities with more precision than is justified by the precision of the original measurements. When *rounding* numbers, first determine the appropriate number of significant digits, then if the next digit (the first *in*significant digit) is less than five, round down. For example, 8.432 rounded to two significant digits is 8.4 because 3 is less than 5. If the first insignificant digit is greater than 5 or 5 followed by anything other than zeros, round up. The value 6.379 with three significant digits rounds to 6.38. If the first insignificant digit is 5 or 5 followed only by zeros, then round up if the last significant digit is odd, and round down if the last significant digit is even. For example, for each of the following, there are two significant digits: 0.2350 rounds to 0.24 (because 3 is odd), 0.2450 rounds to 0.24 (4 is even). If 6.345 is to be rounded to three significant digits, it would become 6.35. If it is to be rounded to two significant digits, it becomes 6.3.[3] (Note that rules vary for rounding when the first insignificant digit is 5 by itself or followed by only zeros. Any reasonable approach to the problem is acceptable, but you should be consistent in the method you use.)

When converting units, round the resulting quantity so that the precision is approximately that of the original measurement. Converting a measurement of 24 feet to meters on a calculator might give: 24 ft · 0.3048 m/ft = 7.3152 m. This should be rounded to 7.3 m because one-half foot, the precision of the original measurement, is comparable to 0.05 m, the precision of the resulting quantity.

When doing arithmetic with measurements, significant digits must be considered. In adding and subtracting, the answer should be rounded so that there are no significant digits further right than the last significant digit on the least precise measurement. Solving 1585.236 + 234.76 + 1.2 gives an answer of 1821.196, which is rounded to 1821.2, because 1.2 has only one decimal place. The answer to 1346.15 − 1218 is 128.15, which is rounded to 128 because 1218 has no decimal places. 23 457 + 9000 + 2800 = 35 257, which is rounded to 35 000 because the last significant digit in 9000 is the 9, which is four places left of the decimal point. If rock A is 1.3 m away from a point, and rock B is 0.12 m further than rock A, then the distance to rock B is 1.3 m + 0.12 m = 1.42 m, which is rounded to 1.4 m, because there is only one digit to the right of the decimal point in 1.3 m. In multiplication and division, the answer should have the same precision as the least precise quantity used. The product of $13.32 \cdot 1.08 \cdot 2$ is 28.7712, which is rounded to 29, because 2 is the least precise value (±0.5). To minimize error, it is best to round numbers after the calculation has been done.

Significant digits are important in science, because they reflect the precision of the measurements made. It is wrong to present data with more digits than are significant because it misleads others into thinking that the measurements are better than they really are, and this affects how the data are interpreted.

## SCIENTIFIC AND PREFIX NOTATION

Large numbers are sometimes shown with scientific and prefix notation, which reduce the number of zeros used. These use exponents, where, for example, $3^4$ means that 3 is multiplied by itself 4 times ($3^4 = 3 \cdot 3 \cdot 3 \cdot 3 = 81$); 4 is called the *exponent*. If $3^4 = 81$, then $3^{-4}$ is the reciprocal of 81 (or 1 ÷ 81 = 0.0123); in this case, −4 is called a *negative exponent*. In *scientific notation*, the base number of 10 is used (e.g., $10^3 = 1000$), and the number is presented with only one digit to the left of the decimal point. For example, 500 can be written as $5 \cdot 10^2$, 500 000 can be written as $5 \cdot 10^5$, and 0.0005 can be written as $5 \cdot 10^{-4}$. Only the significant digits are used in scientific notation. Other examples of scientific notation are 8 650 becomes $8.65 \cdot 10^3$, 9 473 000 becomes $9.473 \cdot 10^6$, and 0.000 499 becomes $4.99 \cdot 10^{-4}$. *Prefix notation* is similar to scientific notation, except that the exponents are always multiples of 3. Because of this, the number of digits left of the decimal point can be one, two, or three. For example, 500 000 becomes $500 \cdot 10^3$, and 0.000 499 becomes $499 \cdot 10^{-6}$. An advantage to prefix notation is that the exponents can be easily converted to SI prefixes, e.g., $499 \cdot 10^{-6}$ m = 499 µm. Keep in mind that $10^6$ is 1 followed by six zeros (1 000 000) and $10^{-6}$ has 1 with five zeros between it and the decimal place (0.000 001).

---

Table A2.1   Rules for exponents (with examples)

| | |
|---|---|
| $10^{-a} = 1 \div 10^a$ | $10^{-4} = 1 \div 10^4 = 1 \div 10\,000 = 0.0001$ |
| $10^a \cdot 10^b = 10^{a+b}$ | $10^2 \cdot 10^3 = 10^{2+3} = 10^5 = 100\,000$ |
| $10^a \div 10^b = 10^{a-b}$ | $10^5 \div 10^2 = 10^{5-2} = 10^3 = 1000$ |
| $10^0 = 1$ | |
| $(10^a)^b = 10^{ab}$ | $(10^3)^2 = 10^{3 \cdot 2} = 10^6 = 1\,000\,000$ |
| $(x \cdot y)^a = x^a \cdot y^a$ | $(5 \cdot 3)^4 = 5^4 \cdot 3^4 = 625 \cdot 81 = 50\,625$ |

---

Multiplying and dividing numbers with exponents are each done in two steps. For example, to find the product of $2.85 \cdot 10^5$ and $8.34 \cdot 10^2$, first solve $2.85 \cdot 8.34$ (= 23.769), then add the exponents ($10^{5+2} = 10^7$). The answer is $23.8 \cdot 10^7$, or 238 000 000, after rounding to three significant digits. To find the quotient of the same two numbers, $2.85 \div 8.34$ (= 0.341 727) and $10^{5-2} = 10^3$, which gives an answer, after rounding, of $0.342 \cdot 10^3$, or 342.

## CALCULATING WITH UNITS

When doing calculations on quantities with units, keep in mind that the units are as important as the numbers. Addition and subtraction can be done only on like units, and the units stay the same; you can add 3 meters and 5 meters to get 8 meters, but you cannot add 3 m to 5 m s$^{-1}$. Multiplication and division can be done on unlike units, and new units are created. For example, if a study site is 50 meters wide and 100 meters long, then the area is $50 \cdot 100 = 5000$ and m $\cdot$ m = m$^2$, or 5000 m$^2$. Traveling 100 m in 20 s gives an average speed of $100 \div 20 = 5$ and m $\div$ s = m s$^{-1}$, or 5 m s$^{-1}$.

Exponents are used with units, as with numbers. For example, if 2 m s$^{-1}$ is squared (2 m s$^{-1}$)$^2$, then the result is 4 m$^2$ s$^{-2}$.

NOTES

1. Paulos, John Allen, 1995, *A Mathematician Reads the Newspaper,* New York: Basic Books, p. 139.

2. Brownridge, Dennis R., 1994, *Metric in Minutes: The Comprehensive Resource for Learning and Teaching the Metric System (SI),* Belmont, California: Professional Publications, Inc., p. 8–15. This reference was used frequently in the preparation of this appendix.

3. This method for rounding is discussed in American National Standards Institute and Institute of Electrical and Electronics Engineers, 1992, *American National Standard for Metric Practice,* ANSI/IEEE Std 268-1992, New York: Institute of Electrical and Electronics Engineers, Inc., p. 61–62.

# WRITING SCIENCE

Good scientific writing is not a matter of life and death; it is much more serious than that.

The goal of scientific research is publication. Scientists, starting as graduate students, are measured primarily not by their dexterity in laboratory manipulations, not by their innate knowledge of either broad or narrow scientific subjects, and certainly not by their wit or charm; they are measured, and become known (or remain unknown) by their publications.

*Robert A. Day, 1994*[1]

Research is not complete until it is presented to the community of scientists. A scientist may put in twenty years of hard work leading to the world's greatest scientific discovery, but if it is not published, it is a waste of effort and time. In the course of doing research, proper notes are of fundamental importance. Writing, then, is a crucial part of the scientific process.

Not all science writing is the same. The majority of papers written by scientists report research projects, and these papers are directed at other experts in a discipline and closely allied fields of study. Review papers are a summary of work done in a particular area, also directed at other researchers. Scientists may also write for a broader audience, explaining aspects of science to the public. An important part of writing is targeting the audience and then using an appropriate style. For example, the terse style of most scientific research papers does not work for popular writing. The focus of this appendix is the scientific research paper, covered after an overview of research notes. More detailed books on writing for scientists are available elsewhere.[2]

Research notebooks were discussed in Chapter 3, so they will be mentioned only briefly here. Laboratory and field notebooks mainly serve as a personal record, but they must be written so that others can understand what was done. Other scientists may need to know details in order to reproduce a study, and historians may want to know how a scientist worked. In the event of charges of misconduct (see Chapter 5), note-books can be used to determine what was actually done in a study. While misconduct charges are relatively rare, all scientists should be pre-pared to defend themselves with proper notebooks. For this reason, notes should be kept in such a way that they cannot be altered. Traditionally, notes were kept in bound books with numbered pages and in permanent ink. In that way, pages cannot be removed or writing erased without leaving evidence of the misdeed.[3] Electronic research notes are used as well but should also meet the requirements of perma-nence and inalterability.[4]

The style of writing for scientific papers, at least the way science *should* be written, is a combination of clarity and brevity. Papers are written to inform others of a research project and, so, should be written in precise and direct language. Journals have a limited number of pages, so to fill them with the most information, papers should not be unnecessarily long. More importantly, scientists are busy people and do not want to spend more time than needed to read a paper. As opposed to creative writing, flowery language is discouraged in science, and important infor-mation should not be revealed only at the end in an attempt to heighten the dramatic tension. (Perhaps flowery language can be used if the paper is about flowers.) Many journals charge the author to publish papers, and the fee is based on the number of pages. That provides an incentive to keep papers short and to the point.

---

Vigorous writing is concise. A sentence should contain no unnecessary words, a paragraph no unnecessary sentences, for the same reason that a drawing should have no unnecessary lines and a machine no unneces-sary parts. This requires not that the writer make all his sentences short, or that he avoid all detail and treat his subjects only in outline, but that every word tell.

William Strunk, Jr., 1918

There you have a short, valuable essay on the nature and beauty of brevity—sixty three words that could change the world.

E. B. White, 1979[5]

---

*Jargon* is the specialized language of a particular group, like police officers or biochemists. The term often is used in a derogatory way, implying unnecessarily obscure words that make it difficult or impossible for outsiders to understand what is being said. Sometimes, though, specialized terms are used because they are more precise than more common words. Take, for example, part of a description of *Artemisia* (sagebrush) shrubs in California: "[Inflorescence] spiciform, racemiform, or paniculate. Heads small, discoid or disciform. . . . Phyllaries dry, imbricate, at least the inner scarious or scarious-margined. Receptacle flat or hemispheric, naked or with long hairs."[6] To most, this sounds like gibberish, but to a botanist it is a very precise and accurate description that would take a long paragraph or two to explain in common language. Whether or not to use jargon depends on the audience. If the audience is other experts who understand the jargon, then it is acceptable to use it. This might be the case in a technical note to a specialized journal. If nonspecialists are expected to read the paper, which is more common, then jargon should be avoided or at least clearly defined. Keep in mind that layers of jargon exist. Where *spiciform* is jargon to nonbotanists, *fractal* is understood by most scientists but is jargon to most nonscientists. Again, writing to a specific audience is important, and choosing appropriate words is a fundamental part of that task. A general rule to follow can be called Occam's razor of science writing: if two words or phrases can say the same thing with equal clarity, use the simpler one; if the simpler wording is considerably longer, then use the one most appropriate for the intended audience. (Occam's razor was discussed in Chapters 8 and 9.) It is wrong to use jargon just to sound impressive or to intentionally make a message obscure!

Scientific papers generally follow the IMRAD approach, with the format of Introduction, Methods, Results, and Discussion. IMRAD works well for most experimental studies but is not appropriate for all scientific reports. A mostly theoretical paper, for example, may have no methods or results to present. The format of any paper is determined by the subject matter. Since IMRAD works for many science papers, it will be discussed here. (Acronyms like IMRAD are jargon, but if explained properly, they can be used effectively.)

Finally, there is the question of clarity. This is largely a question of psychology. Does the writer really want to convey information to his readers, or is he trying to impress them with his own genius? Unfortunately, some scientists suffer from an inferiority complex which continually compels them to bolster their egos by writing papers so obscure than even the most brilliant specialists in the same field cannot understand them. What a triumph!

E. Bright Wilson, Jr., 1952[7]

The purpose of the *introduction* is to state what the subject of the study is, why it was done, and to put it in context with previous work on the topic. For example, a paper might start with "Microeconomic theory suggests that small business owners will. . . . In this study, we interviewed bookstore owners to see if they follow the general trend." The rest of the introduction would then explain the background of that theory, describing previous relevant work in the area, then summarize what was done in this study, and finally explain why this is important from a theoretical or applied (or both) perspective.

The *methods* section is where the experimental design is presented. Reproducibility is of fundamental importance to science, so the methods must be explained in sufficient detail that a competent scientist in that field can perform the same experiment. Some methods are standardized and need little explanation (we weighed the samples to the nearest gram . . .). However, any new or modified instruments, experimental methods, or analysis procedures will have to be explained in detail. Likewise, some materials may have to be described explicitly. Water can be distilled or from the tap, for example, and if that difference can affect the outcome of the experiment, the form of water must be specified. Since no measurement in science is perfect, as discussed in the previous appendix, the precision and accuracy of the methods should be explained clearly.

In the *results* section, the data collected are presented in a meaningful way. Typically, not all of the data are given, just those relevant to the paper and usually in a summarized form. For example, 100 measurements may be collected, but presented only as a mean value and standard deviation. Tables and graphs (see Appendix 4) are often used to help the reader see any relevant patterns. These patterns are then described in words, as in "The population increased logarithmically over time, as shown in Figure 3."

Where the data are presented and described in the results section, they are interpreted in the *discussion* section. Here, various questions can be addressed. What generalizations can be made from the results? Do they agree with the results of previous studies? What are the theoretical implications and practical applications of the knowledge gained from the study? What are the strengths and weaknesses of the study? Discussing the weakness of the research is important in science writing. The scientist doing the research usually is best qualified to comment on its problems. The purpose of the discussion is to suggest to other scientists what they should learn from the study, and it is wrong to paint an unreasonably rosy picture of the work.

Often presented as part of the discussion section, the *conclusions* are a summary of the main findings of the study. Here the reader is reminded, in brief, what is important about the study. This can include theoretical contributions, improved experimental methods, and insight gained on what research is needed to further knowledge in the subject area.

Other parts of a scientific paper are the abstract, keywords, acknowledgments, and references. The *abstract* is a brief summary of the entire paper, put at the beginning. Most scientists will read an abstract first and then decide if they want to read the paper itself. After the abstract is a list of *keywords*, words or short phrases describing the main topics of the paper (e.g., Darwin's Finches, speciation, Galapagos Islands). In the *acknowledgments*, people who helped the research are thanked and sources of funding are noted. All published works mentioned in the text are listed in the *references* section.

Writing is an important skill for scientists to develop. Inability to write effectively diminishes a scientist's contribution to knowledge and can hinder a career. Another aspect of scientific communication is graphing, the subject of the next appendix.

## NOTES

1. Day, Robert A., 1994, *How to Write and Publish a Scientific Paper*, 4th ed., Phoenix, Arizona: Oryx Press, p. ix.

2. For example, see the following: O'Connor, Maeve, 1991, *Writing Successfully in Science*, London: HarperCollins Academic; Shortland, Michael, and Gregory, Jane, 1991, *Communicating Science: A Handbook*, Essex, UK: Longman Scientific and Technical; Day, 1994; Porush, David, 1995, *A Short Guide to Writing About Science*, New York: Harper-Collins College Publishers; Moriarty, Marilyn F., 1997, *Writing Science Through Critical Thinking*, Boston: Jones and Bartlett Publishers.

3. Kanare, Howard M., 1985, *Writing the Laboratory Notebook*, Washington, D.C.: American Chemical Society.

4. See, for example, U.S. Environmental Protection Agency, Office of Information Resources Management, Scientific Systems Staff, 1995 edition, Directive 2185: *Good Automated Laboratory Practices: Principles and Guidance to Regulations for Ensuring Data Integrity in Automated Laboratory Operations with Implementation Guidance*. Research Triangle Park, North Carolina: U.S. EPA Office of Information Resources Management. This document can be found at http://www.epa.gov/irmpoli8/irm_galp/index.html. Also, see Food and Drug Administration, Department of Health and Human Services, 1997, "Electronic Records; Electronic Signatures," *Federal Register* 62, no. 54: 13429–66.

5. Strunk, William, and White, E. B., 1979, *The Elements of Style*, 3rd ed., New York: Macmillan Publishers, p. xiv. White revised and updated Strunk's book, originally published in 1918.

6. Munz, Phillip A., 1974, *A Flora of Southern California*, Berkeley: University of California Press, p. 116.

7. Wilson, E. Bright, Jr., 1952, *An Introduction to Scientific Research*, New York: McGraw-Hill Book Company, p. 357.

# GRAPHING

Modern data graphics can do much more than simply substitute for small statistical tables. At their best, graphics are instruments for reasoning about quantitative information. Often the most effective way to describe, explore, and summarize a set of numbers—even a very large set—is to look at pictures of those numbers. Furthermore, of all methods for analyzing and communicating statistical information, well-designed data graphics are usually the simplest and at the same time the most powerful.

*Edward R. Tufte, 1983*[1]

An important part of scientific communication is the presentation of data, often in the form of graphs. Graphing is used in the data analysis stage of science, where graphs are used to search for patterns in the data collected. They are also used in the presentation of research results to others in, for example, journal articles and oral presentations. In either case, a well-designed graph is important. This appendix is a quick introduction to the creation of simple graphs, emphasizing those used to present information to others.[2]

Graphs vary considerably in their look but share some common elements. Most are rectangular in shape and will have an X-axis on the bottom and a Y-axis on the left side. Each axis has tick marks and numbers showing the values along it, and each has a label to identify what is being represented by that axis and the units used. The content of the graph is the information (symbols, lines, etc.) inside the box, showing the actual data and other relevant information. Graphs typically have a title to tell what they are about. The title may be at the top of the graph or given as the first sentence of a caption. Captions often include additional information useful in interpreting the data.

There are two main aspects of graphing: accuracy and design. The data must be portrayed correctly and clearly so that the user of the graph readily understands the relationship shown. Graphs can be drawn by hand, but typically they are created with graphing software. In computer generated graphs, accuracy, meaning data points placed in the correct location with respect to the axes, is not a problem as long as the data are entered properly. It is the design of graphs that requires the most effort and skill. Graph design involves an assessment of the following questions: What is the best type of graph to show the data? Who is the audience? What are they expected to learn from the graph? and How is it going to be seen?

With regard to the first question, different types of data need to be displayed in different ways. The relationship between two variables is shown in a *scatterplot*. When one variable is continuous, such as time, a *line graph* is used. A *bar graph* is used when one of the axes represents categories instead of quantitative data, and a *histogram* shows the distribution of a single variable. Each of these will be discussed. Other forms of graphical display of data, such as pie charts and maps, will not be discussed here, though many of the design concepts apply to them as well.

The questions "Who is the audience?" and "What are they expected to get from the graph?" determine how the data are portrayed within a particular format. If the data show how height varies with age, a scatterplot for children to use might have only the data points themselves, while a graph of the same data set for researchers in pediatrics might have the data points, an equation for the curve that best describes the trend of the data, error bars to show the variation in values represented by each data point, and a grid to aid in determining the precise values of the data points. In addition, the graph for scientists might have different symbols to represent subsets of the data, such as gender or socioeconomic background of the children measured. Making the graph too complicated for the audience detracts from its usefulness, as does leaving out valuable detail. Targeting the audience and intended use are crucial in preparing to make a graph.

Where the graph is going to be seen is also an important factor in design. A graph in a journal article or book can be looked at as long as the reader chooses, so it can be reasonably detailed. A graph projected on a screen as part of an oral presentation, however, may be seen only for a minute or less; such a graph needs to be designed to be easily understandable (few details) and readable from the back of the room (big numbers and symbols).

As stated previously, the purpose of *scatterplots* is to show the relationship between two variables. (They can show more than two variables, but this is a simplified introduction to graphing.) A study on the relationship between fox and rabbit population densities in a particular area might involve measuring the populations at various times over a period of several years. A scatterplot then could be prepared where each data point shows both populations at a given time, as shown in Figure A4.1a. On graphs the X-axis, by convention, is used for the *independent variable,* the variable which is not affected by the other. The Y-axis, then, is for the *dependent variable*, the one whose value is influenced by the independent variable, but this is not always the case. With foxes and rabbits, there may be no real independent variable, since neither is independent of the other—rabbits provide food for foxes, and foxes kill rabbits. So the general rule is: if there is an independent variable, it is put on the X-axis. Figure A4.1a on page 175 shows that as fox density increases, rabbit density increases as well in a roughly linear fashion. The same data are plotted in Figure A4.1b, but the range of the Y-axis has been changed. A quick look at the two graphs suggests that the data points in Figure A4.1a are more scattered (less lined up) than in Figure A4.1b, and therefore, the impression is given in Figure A4.1b that the relationship between foxes and rabbits is stronger. However, it is important to remember than they are the same data points. In designing graphs, it is improper to pick the axis range so that the data "look better" than they really are. In reading graphs, it is useful to think about how the design of the graph affects the image it gives of the data.

The same data are plotted in Figure A4.1c but with more information included. If the data points on a scatterplot form a pattern of some sort, then a curve can be fit to show the trend of the relationship, as done on Figure A4.1c. For example, if the pattern of the points is linear, as it is here, then a straight line is fit through the points, which best shows the trend of those points. (Yes, a straight line is still called a curve.) The equation at the top of the graph describes the line and can be used to calculate the expected rabbit density for any fox density. One measure of the degree to which those points follow that line is called the correlation coefficient (symbol $r$) so that if all the points were exactly on the line, $r$ would be 1.0 (or –1.0 if the line sloped downward), and if there was no pattern at all (the distribution of points was purely random), then $r$ would be 0.0. In other words, $r = 0.86$ means that the line better represents that relationship than if $r = 0.43$. (For details on curve fitting and correlation coefficients, see a statistics text.) While Figures A4.1a and A4.1b give different impressions of the degree of correlation of the two variables, the correlation coefficient (and other similar coefficients not discussed here) is an objective measure, not dependent on the way the data are portrayed.

Also shown in Figure A4.1c are *error bars*, the lines radiating from each point. Since no measurement is perfect, as discussed in Appendix 2, no data point on a graph is perfect either. In the hypothetical case of the foxes and rabbits, an estimate was made of the accuracy of the field methods used to estimate the populations (since not all animals cooperate when being counted). The points are plotted in the most likely position, but the error bars show that the actual values may be significantly different. If the points plotted represent averaged values of many measurements, then the error bars can represent a statistical measure of how much the individual values vary from the average (such as standard deviation and standard error—see a statistics text for details). What the error bars represent should be clearly stated in the caption of the graph. Depending on the situation, error bars can be used for the X-axis, the Y-axis, or both, as in Figure A4.1c. If, for example, rabbit populations could be counted with great precision, then the vertical error bars would not be needed. Not showing error bars when the imprecision of the measurements is significant misleads the reader. On the other hand, for graphs made just to show general trends, error bars may add unnecessary clutter. It depends on the intended use of the graph. Also keep in mind that error bars are not limited to scatterplots; they can be used on any type of graph. Figure A4.1d shows the same data as Figure A4.1c but with grid lines added. These help the reader make accurate estimations of the values of each point. They also add to the clutter of the graph and should be used only when needed. Graphs should be made complicated only

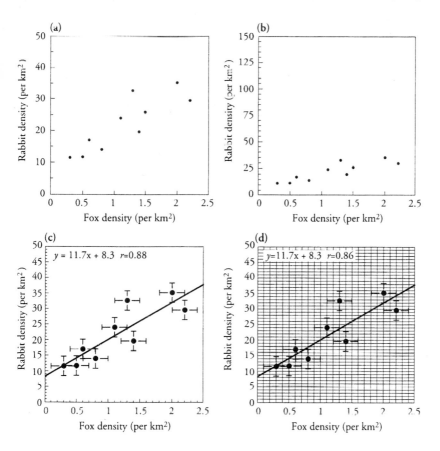

Figure A4.1   Fox and rabbit population densities for Endeavor Island. Each point represents population estimates made in July from 1990 to 1999. Error bars represent estimated error in sampling methods, 0.2 per km² for foxes and 3 per km² for rabbits. (Hypothetical Data.)

when there is a good reason for doing so. Other examples of scatterplots are graphs of the amount of sleep and a measure of worker productivity, and of the damage from a hurricane with distance from the coast.

While scatterplots show discrete data points, *line graphs* are used when the independent variable is continuous and the purpose of the graph is to show how the dependent variable changes with the independent one. Figure A4.2a shows average monthly temperatures for a year. Time is the independent variable (change in temperature has no effect on time) and

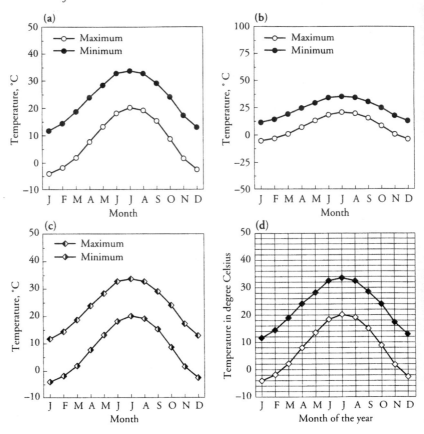

Figure A4.2   Averaged maximum and minimum monthly temperatures for Lubbock, Texas. (*Source:* 1997 Local Climatological Data Annual Summary with Comparative Data, Lubbock, Texas, National Climatic Data Center, Asheville, North Carolina.)

so is plotted on the X-axis. Temperature, then, is put on the Y-axis. This is a *time series* graph, one of the most common types of line graph. There is one data point for each month, and these points are connected by lines. The implication is that temperature changes evenly along the lines between the points. This is a reasonable assumption with averaged temperatures, which do not vary much over days or weeks. However, if the graph were of actual temperatures, not averaged ones, then there would be much variation over short time periods, and the lines would not reflect the actual change. Part of experimental design is to pick a measurement interval that provides sufficient detail on the real fluctuations in a measured quantity. For example, hourly measurements generally will reflect the actual variation in air temperature. It is not always feasible to measure at the appropriate interval; in such situations, it is the responsibility of the scientist to state clearly the limitations of the data and that the line on the graph may not accurately reflect the true variation.

Several other versions of Figure A4.2a are also presented. Figure A4.2b is another example of changing the look of the data by changing the range of the Y-axis. As a general guideline, Figure A4.2a is better because the data fill up most of the area of the graph. If, however, this graph must be compared to other graphs of temperature at other locations, then the Y-axis would be the same for all of the graphs, and the range would be chosen to include the highest and lowest values on all of the data. A legend has been included on these graphs to indicate what each symbol represents. (Which symbol represents maximum and which represents minimum is fairly obvious on these graphs, so the legend may not be needed, but often it is helpful on graphs.) The symbols have been changed on Figure A4.2c. This is to show that selection of symbols is important—the reader of the graph should not have to work to distinguish the difference between them. Finally, in Figure A4.2d, the grid lines obscure the data points, and the axis labels are longer than they need to be. Again, simplicity and ease of use are desirable. If grid lines are deemed to be needed, then Figure A4.2d could be improved by reducing the line width of the grid and increasing the size of the symbols to make them more visible.

Other examples of line graphs are graphs of imports and exports for a country over time and change in air pressure with elevation. Just to show that rules can be broken, in the latter graph, elevation would be plotted on the Y-axis even though it is the independent variable. The vertical phenomenon of elevation just does not look right when presented horizontally. Conveying the correct message is more important than blindly following a rule.

*Bar graphs* are used to show relative values of distinct entities. One of the axes is not quantitative, it just allows the categories to be identified. A graph of the human population of each continent (Figure A4.3a) uses the X-axis to identify the continent and the Y-axis to show the popula-

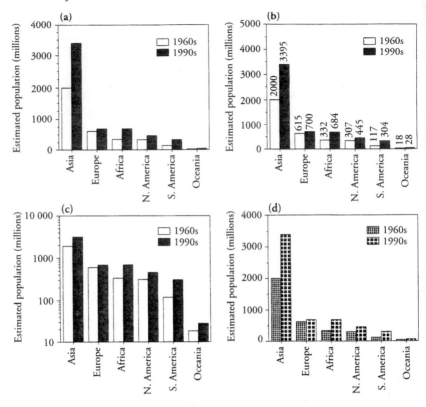

Figure A4.3   Continental populations in the 1960s and 1990s. (*Source: Goode's World Atlas*, 13th ed., 1970 and 19th ed., 1995. Map © by Rand McNally R.L. #99-S-59.)

tion. The bars have equal width and vary only in their height. Sometimes all of the information cannot be shown clearly on a graph, as shown in the data for Oceania (Australia and the Pacific Islands) in Figure A4.3a. Here the population is so low that the bars barely show. One way around this problem is to add the values for each bar in the graph, as in Figure A4.3b. This is also done on bar graphs when precise values are needed by the reader. Another approach when there are very small values and very large ones is to make the Y-axis logarithmic rather than arithmetic, as in Figure A4.3c, but that would make it more difficult to see the amount of change between the decades. (Logarithmic axes are useful and common in science but not appropriate in this case.)

In graph design, compromises must be made, with the approach used chosen because it works best for the intended audience. Figure A4.3d is shown to emphasize that there is an aesthetic aspect of graph design, and annoying patterns are rarely justified. Examples of other data plotted with bar graphs are the amount of improvement in a group of patients given a new drug and another group given a placebo (a bar for drug and another for placebo), chemical reaction times for different solutions (a bar for each solution used), or average wages for males and females in a company (one bar for men, one for women).

A *histogram* is similar to a bar graph, except that both axes are quantitative and distributions of one variable are portrayed. Histograms typically are used to show the distribution of some thing. An example is the distribution of the populations of the states in the United States, as shown in Figure A4.4a. The X-axis shows the population categories, and the Y-axis is the number of states in each category. In this case, eight states have between 0 and 1 million people, while only one (California, if you are interested) has between 31 and 32 million. The size of the categories is an important consideration in histogram design. The four examples show the same data but with different categories and different Y-axis ranges. Which would be used depends on the intended message of the graph. Figure A4.4a may be too detailed and Figure A4.4d too general to be of much use. Of course, it depends on the audience and what they are expected to learn from the graph. Other examples of histograms are the distributions of test scores for an exam, earthquake intensities in Japan, and salaries of corporate executive officers.

There is an artistic aspect of graphing, which is important. Scientists do develop their own styles in terms of such items as type fonts and symbols used. Some even like to get fancy, with the graphs superimposed on background images and imaginative color schemes. These can be effective, especially during oral presentations, as long as the message of the graph is not lost in the excitement of the visual display. Message first, beauty second.

Designing graphs is a skill that scientists should work on, just like gaining expertise in research methods and in writing. Much trial and error is involved in making an effective graph, but the results are usually

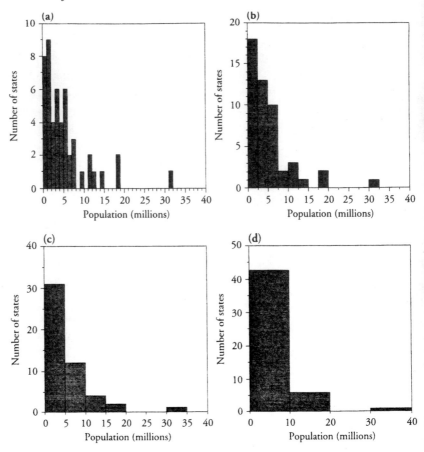

Figure A4.4    Distribution of state populations.

worth the effort. The value of scientific work is diminished if it is not communicated effectively.

## NOTES

1. Tufte, Edward R., 1983, *The Visual Display of Quantitative Information*, Cheshire, Connecticut: Graphics Press, p. 9.

2. More detailed discussions of graphing can be found in Cleveland, William S., 1985, *The Elements of Graphing*, Monterey, California: Wadsworth Advanced Books and Software; Kosslyn, Stephen, 1994, *Elements of Graph Design*, New York: W. H. Freeman; and Dent, Borden D., 1999, *Cartography: Thematic Map Design*, 5th ed., Boston: WCB McGraw-Hill, p. 358–84.

# CREDITS

Page 6: Excerpt from Carl Sagan, *The Demon-Haunted World: Science as a Candle in the Dark* (NY: Random House, 1995). Reprinted by permission of Random House, Inc.

Page 56: Excerpt from Oliver la Farge, "Scientists are lonely men," *Harper's Magazine* (1942) page 652.

Pages 103–104: Excerpt reprinted with permission from Shawn Carlson, "A double-blind test of astrology," *Nature* (1985) 318:425. ©1985 Macmillan Magazines Ltd.

Page 109: "Wisdom of dialectical materialism" from *Albert Einstein: Creator and Rebel* by Banesh Hoffman with Helen Dukas. Copyright ©1972 by Helen Dukas and Banesh Hoffman. Used by permission of Dutton, a division of Penguin Putnam, Inc.

Pages 113–114: Excerpt from Jane Ayers Sweat and Mark Durm, "Psychics: do police departments really use them?" *Skeptical Enquirer* (1993) 17: 156-157. Used by permission of the Skeptical Enquirer.

Page 114: Poem translation in *Nostradamus on Napoleon, Hitler and the Present Crisis* by Stewart Robb (NY: Charles Scribner's Sons, 1941).

Page 118: Excerpt from Phil B. Fontanarosa and George Lunberg, "Alternative medicine meets science," *JAMA* (1998) 280:1619.

Page 130: Reprinted with the permission of The Free Press, a division of Simon & Schuster, Inc., from *How We Know What Isn't So: The Fallibility of Human Reason in Everyday Life*. Copyright ©1991 by Thomas Gilovich.

Page 131: Excerpts from Mario Bunge, "A skeptic's beliefs and disbeliefs," *New Ideas in Psychology* (1991) 9:131.

Page 139: Excerpt from Plato, *The Republic*, Book 10, translated by Benjamin Jowett (Cleveland: The World Publishing Company, 1946), by permission of Oxford University Press.

# INDEX